Hall-Effect Sensors

Hall-Effect Sensors
Theory and Applications

by Edward Ramsden

ELSEVIER

AMSTERDAM • BOSTON • HEIDELBERG • LONDON
NEW YORK • OXFORD • PARIS • SAN DIEGO
SAN FRANCISCO • SINGAPORE • SYDNEY • TOKYO
Newnes is an imprint of Elsevier

Newnes

Newnes is an imprint of Elsevier
30 Corporate Drive, Suite 400, Burlington, MA 01803, USA
Linacre House, Jordan Hill, Oxford OX2 8DP, UK

 Recognizing the importance of preserving what has been written, Elsevier prints its
books on acid-free paper whenever possible.

Library of Congress Cataloging-in-Publication Data

Application submitted

British Library Cataloguing-in-Publication Data
A catalogue record for this book is available from the British Library.

ISBN-13: 978-0-7506-7934-3

For information on all Newnes publications
visit our website at www.books.elsevier.com.

Transferred to Digital Printing 2009

Contents

Chapter 3: Transducer Interfacing 35

Chapter 4: Integrated Sensors: Linear and Digital Devices 61

Chapter 5: Interfacing to IntegratedHall-Effect Devices...............83

Introduction

Among the various sensing technologies used to detect magnetic fields, the Hall effect is perhaps the most widespread and commonly used. Because it is possible to construct high-quality Hall-effect transducers with the standard integrated circuit processes used in the microelectronics industry, and to integrate ancillary signal-processing circuitry on the same silicon die, usable sensors can be fabricated readily and inexpensively. Hundreds of millions of these devices are produced every year for use in a wide variety of applications. A few places where Hall-effect transducers can be found are:

- *Automobiles* – Ignition timing, antilock braking (ABS) systems,
- *Computers* – Commutation for brushless fans, disk drive index sensors,
- *Industrial Controls* – Speed sensors, end-of-travel sensors, encoders,
- *Consumer Devices* – Exercise equipment, cell phones.

Knowledge of the Hall effect itself is quite old, as it was discovered experimentally by Edwin Hall in 1879. The discovery of the effect, incidentally, preceded the discovery of the electron by Thomson in 1897 by nearly 20 years. At the time Hall was performing his experiments, electric current was commonly believed to be a continuous fluid, not a collection of discrete elementary particles.

It is quite remarkable that Hall's experiments allowed him to observe the effect at all, when one considers the instrumentation available at the time and the subtle nature of the experiment, which most likely provided signals of only microvolts. Nevertheless, the Hall effect became reasonably well known early on; the Smithsonian Institute Physical Tables from 1920 include a table describing the magnitude of the Hall effect for a number of substances [Fowle20].

In the 1950s, Hall-effect transducers were commonly used to make laboratory-type magnetic measurement instruments. The availability of semiconductor materials enabled the fabrication of high-quality transducers.

In the 1960s and 1970s, it became possible to build Hall-effect sensors on integrated circuits with on-board signal-processing circuitry. This advance vastly reduced the cost of using these devices, enabling their widespread practical use. One of the first major applications was in computer keyboards, where the new integrated Hall-effect sensors were used to replace mechanical contacts in the key switches. By substituting solid-state sensors for electro-mechanical contacts, the reliability and durability of keyboards were vastly improved.

As it became possible to put more transistors on a given-sized silicon die, the basic transducer could be surrounded with support functions at little additional cost. This allowed Hall-effect sensors with on-board logic for bus interfacing, temperature compensation, and application-specific signal processing. Hall-effect ICs with sophisticated on-chip interface circuitry began appearing in the late 1980s, with new devices still being developed to meet the needs of specialized applications.

While the end-objective of measuring a magnetic field is rare outside of a physics laboratory, magnetic fields make useful intermediaries for sensing other phenomena. Because large magnetic fields are not commonly encountered in nature and can pass through most materials unhindered, they make flexible and vivid indicators when they can be controlled by other phenomena. One simple example of this is in proximity detection, which is the function of sensing if an object is present or absent. The characteristics of the object may make it difficult to directly sense its presence in a given environment. An attached magnet, however, can make it very easy to detect under a variety of conditions. While the ultimate goal is the detection of the object, it is accomplished in this case by the detection of a magnetic field. The most common sensing applications for Hall-effect sensors are proximity, position, speed, and current. Integrated Hall-effect sensors are the preferred choice for a number of reasons:

- *Small Size* – Integrated Hall-effect sensors with on-board amplifiers can be obtained in surface-mount IC packages, taking up no more area on a printed circuit board than a discrete transistor. Simple Hall-effect transducers can be obtained in packages that are nearly microscopic. The small size of Hall-effect sensors allows them to physically fit into many places where other magnetic transducers would be too bulky.

- *Ruggedness* – Because most Hall-effect sensors are fabricated as monolithic integrated circuits, they are very immune to shock and vibration. In addition, standard IC packaging is highly resistant to moisture and environmental contaminants. Finally, monolithic Hall-effect ICs that operate over the temperature range of –40°C to +150°C are readily available from a number of sources. Hall-effect ICs have been successfully used in hostile environments such as inside automotive transmissions and down the bore-hole in oil-well drilling equipment.

- *Ease-of-Use* – While Hall-effect transducers are not even close to being the most sensitive or accurate means of measuring magnetic fields, they are predictable and well-behaved. The output of a Hall-effect transducer is nearly linear over a substantial range of magnetic field and exhibits no significant hysteresis or memory effects. Unlike many types of magnetic sensors, Hall-effect transducers can differentiate north and south fields. Because of their small size, they are effectively "point" sensors, measuring the field at a single point in space. Finally,

Hall-effect transducers measure a single spatial component of a field, allowing one to sense the direction of a field as well as its magnitude.

- *Cost* – While an instrumentation-grade Hall-effect sensor can cost several hundred dollars, the vast majority of transducers currently produced in the world are sold for less than 50 cents, including signal-processing electronics. Hall-effect sensors are among the most cost-effective magnetic field sensors available today.

For all of the previous reasons, Hall-effect sensors are useful items in the system designer's toolbox. Subsequent chapters of this book will explain how they work, how to interface with them, and how to use them in real-world applications.

Chapter 1

Hall-Effect Physics

Conceptually, a demonstration of the Hall effect is simple to set up and is illustrated in Figure 1-1. Figure 1-1a shows a thin plate of conductive material, such as copper, that is carrying a current (I), in this case supplied by a battery. One can position a pair of probes connected to a voltmeter opposite each other along the sides of this plate such that the measured voltage is zero.

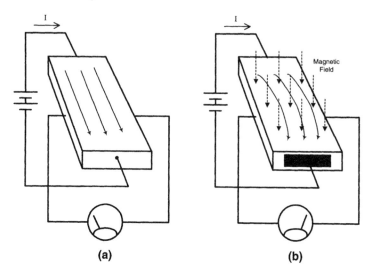

(a)   **(b)**

Figure 1-1: The Hall effect in a conductive sheet.

When a magnetic field is applied to the plate so that it is at right angles to the current flow, as shown in Figure 1-1b, a small voltage appears across the plate, which can be measured by the probes. If you reverse the direction (polarity) of the magnetic field, the polarity of this induced voltage will also reverse. This phenomenon is called the Hall effect, named after Edwin Hall.

What made the Hall effect a surprising discovery for its time (1879) is that it occurs under steady-state conditions, meaning that the voltage across the plate persists even when the current and magnetic field are constant over time. When a magnetic field varies with time, voltages are established by the mechanism of induction, and induction was well understood in the late 19th century. Observing a short voltage pulse across the plate when a magnet was brought up to it, and another one when the magnetic field was removed, would not have surprised a physicist of that era. The continuous behavior of the Hall-effect, however, presented a genuinely new phenomenon.

Under most conditions the Hall-effect voltage in metals is extremely small and difficult to measure and is not something that would likely have been discovered by accident. The initial observation that led to discovery of the Hall effect occurred in the 1820s, when Andre A. Ampere discovered that current-carrying wires experienced mechanical force when placed in a magnetic field (Figure 1-2). Hall's question was whether it was the wires or the current in the wires that was experiencing the force. Hall reasoned that if the force was acting on the current itself, it should crowd the current to one side of the wire. In addition to producing a force, this crowding of the current should also cause a slight, but measurable, voltage across the wire.

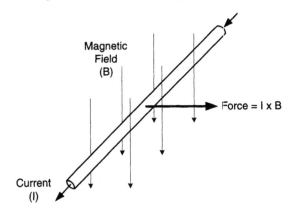

Figure 1-2: A magnetic field exerts mechanical force on a current-carrying wire.

Hall's hypothesis was substantially correct; current flowing down a wire in a magnetic field does slightly crowd to one side, as illustrated in Figure 1-1b, the degree of crowding being highly exaggerated. This phenomenon would occur whether or not the current consists of large numbers of discrete particles, as is now known, or whether it is a continuous fluid, as was commonly believed in Hall's time.

1.1 A Quantitative Examination

Enough is presently known about both electromagnetics and the properties of various materials to enable one to analyze and design practical magnetic transducers based on the Hall effect. Where the previous section described the Hall effect qualitatively, this section will attempt to provide a more quantitative description of the effect and to relate it to fundamental electromagnetic theory.

In order to understand the Hall effect, one must understand how charged particles, such as electrons, move in response to electric and magnetic fields. The force exerted on a charged particle by an electromagnetic field is described by:

$$\vec{F} = q_0 \vec{E} + q_0 \vec{v} \times \vec{B}$$

(**Equation 1-1**)

where $\vec{F}$ is the resultant force, $\vec{E}$ is the electric field, $\vec{v}$ is the velocity of the charge, $\vec{B}$ is the magnetic field, and q_0 is the magnitude of the charge. This relationship is commonly referred to as the Lorentz force equation. Note that, except for q_0, all of these variables are vector quantities, meaning that they contain independent x, y, and z components. This equation represents two separate effects: the response of a charge to an electric field, and the response of a moving charge to a magnetic field.

In the case of the electric field, a charge will experience a force in the direction of the field, proportional both to the magnitude of the charge and the strength of the field. This effect is what causes an electric current to flow. Electrons in a conductor are pulled along by the electric field developed by differences in potential (voltage) at different points.

In the case of the magnetic field, a charged particle doesn't experience any force unless it is moving. When it is moving, the force experienced by a charged particle is a function of its charge, the direction in which it is moving, and the orientation of the magnetic field it is moving through. Note that particles with opposite charges will experience force in opposite directions; the signs of all variables are significant. In the simple case where the velocity is at right angles to the magnetic field, the force exerted is at right angles to both the velocity and the magnetic field. The cross-product operator ($\times$) describes this relationship exactly. Expanded out, the force in each axis (x,y,z) is related to the velocity and magnetic field components in the various axes by:

$$F_x = q_0 \left(v_y B_z - v_z B_y \right)$$
$$F_y = q_0 \left(v_z B_z - v_z B_x \right)$$
$$F_z = q_0 \left(v_x B_y - v_y B_x \right)$$

(**Equation 1-2**)

The forces a moving charge experiences in a magnetic field cause it to move in curved paths, as depicted in Figure 1-3. Depending on the relationship of the velocity to the magnetic field, the motion can be in circular or helical patterns.

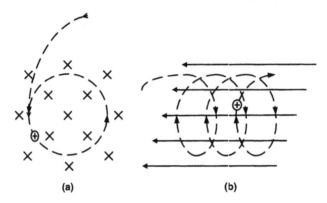

(a)   (b)

Figure 1-3: Magnetic fields cause charged particles to move in circular (a) or helical (b) paths.

In the case of charge carriers moving through a Hall transducer, the charge carrier velocity is substantially in one direction along the length of the device, as shown in Figure 1-4, and the sense electrodes are connected along a perpendicular axis across the width. By constraining the carrier velocity to the x axis ($vy = 0$, $vz = 0$) and the sensing of charge imbalance to the z axis, we can simplify the above three sets of equations to one:

$$F_z = q_0 v_x B_y$$ **(Equation 1-3)**

which implies that the Hall-effect transducer will be sensitive only to the y component of the magnetic field. This would lead one to expect that a Hall-effect transducer would be orientation sensitive, and this is indeed the case. Practical devices are sensitive to magnetic field components along a single axis and are substantially insensitive to those components on the two remaining axes. (See Figure 1-4.)

Although the magnetic field forces the charge carriers to one side of the Hall transducer, this process is self-limiting, because the excess concentration of charges to one side and consequent depletion on the other gives rise to an electric field across the transducer. This field causes the carriers to try to redistribute themselves more evenly. It also gives rise to a voltage that can be measured across the plate. An equilibrium develops where the magnetic force pushing the charge carriers aside is balanced out by the electric force trying to push them back toward the middle

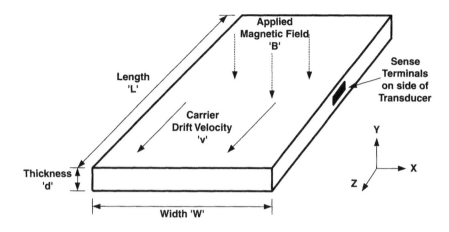

Figure 1-4: Hall-effect transducer showing critical dimensions and reference axis.

$$q_0 E_H + q_0 v \times B = 0 \qquad \textbf{(Equation 1-4)}$$

where E_H is the Hall electric field across the transducer. Solving for E_H yields

$$E_H = -v \times B \qquad \textbf{(Equation 1-5)}$$

which means that the Hall field is solely a function of the velocity of the charge carriers and the strength of the magnetic field. For a transducer with a given width w between sense electrodes, the Hall electric field can be integrated over w, assuming it is uniform, giving us the Hall voltage.

$$V_H = -wvB \qquad \textbf{(Equation 1-6)}$$

The Hall voltage is therefore a linear function of:
    a) the charge carrier velocity in the body of the transducer,
    b) the applied magnetic field in the "sensitive" axis,
    c) the spatial separation of the sense contacts, at right angles to carrier motion.

1.2 Hall Effect in Metals

To estimate the sensitivity of a given Hall transducer, it is necessary to know the average charge carrier velocity. In a metal, conduction electrons are free to move about and do so at random because of their thermal energy. These random "thermal velocities" can be quite high for any given electron, but because the motion is random, the motions of individual electrons average out to a zero net motion, resulting in no current. When an electric field is applied to a conductor, the electrons "drift" in the direction of the applied field, while still performing a fast random walk from their thermal energy. This average rate of motion from an electric field is known as *drift velocity*.

In the case of highly conductive metals, drift velocity can be estimated. The first step is to calculate the density of carriers per unit volume. In the case of a metal such as copper, it can be assumed that every copper atom has one electron in its outer shell that is available for conducting electric current. The volumetric carrier density is therefore the product of the number of atoms per unit of weight and the specific gravity. For the case of copper this can be calculated:

$$N = \frac{N_A}{M_m}D = \frac{6.02 \times 10^{23} \text{ mol}^{-1}}{63.55 \text{g} \cdot \text{mol}^{-1}} \times 8.89 \text{g} \cdot \text{cm}^{-3} = 8.42 \times 10^{22} \text{ cm}^{-3} \quad \textbf{(Equation 1-7)}$$

where: N is the number of carriers per cubic centimeter
        N_A is the Avogadro constant (6.02×10^{23} mol^{-1})
        M_m is the molar mass of copper (63.55 g $\cdot$ mol$^-$)
        D is specific gravity of copper (grams/cm3)

Once one has the carrier density, one can estimate the carrier drift velocity based on current. The unit of current, the ampere (A), is defined as the passage of $\approx 6.2 \times 10^{18}$ charge carriers per second and is equal to $1/q0$. Consider the case of a piece of conductive material with a given cross-sectional area of A. The carrier velocity will be proportional to the current, as twice as much current will push twice as many carriers through per unit time. Assuming that the carrier density is constant and the carriers behave like an incompressible fluid, the velocity will also be inversely proportional to the cross section, a larger cross section meaning lower carrier velocity. The carrier drift velocity can be determined by:

$$v = \frac{I}{q_0 NA} \qquad\qquad\qquad\qquad \textbf{(Equation 1-8)}$$

where
        v is carrier velocity, cm/sec
        I is current in amperes
        Q_0 is the charge on an electron (1.60×10^{-19} C)
        N is the carrier density, carriers/cm^3
        A is the cross section in cm^2

One surprising result is the drift velocity of carriers in metals. While the electric field that causes the charge carriers to move propagates through a conductor at approximately half the speed of light (300×10^6 m/s), the actual carriers move along at a much more leisurely average pace. To get an idea of the disparity, consider a piece of

#18 gauge copper wire carrying one ampere. This gauge of wire is commonly used for wiring lamps and other household appliances and has a cross section of about 0.0078 cm². One ampere is about the amount of current required to light a 100-watt light bulb. Using the previously derived carrier density for copper and substituting into the previous equation gives:

$$v = \frac{1A}{1.6 \times 10^{-19} C \cdot 8.42 \times 10^{22} cm^{-3} \cdot 0.0078 \, cm^2} = 0.009 \, cm \cdot s^{-1} \quad \textbf{(Equation 1-9)}$$

The carrier drift velocity in the above example is considerably slower than the speed of light; in fact, it is considerably slower than the speed of your average garden snail.

By combining Equations (1-6) and (1-8), we can derive an expression that describes the sensitivity of a Hall transducer as a function of cross-sectional dimensions, current, and carrier density:

$$V_H = \frac{IB}{q_0 N d} \quad \textbf{(Equation 1-10)}$$

where d is the thickness of the conductor.

Consider the case of a transducer consisting of a piece of copper foil, similar to that shown back in Figure 1-1. Assume the current to be 1 ampere and the thickness to be 25 µm (0.001"). For a magnetic field of 1 tesla (10,000 gauss) the resulting Hall voltage will be:

$$V_H = \frac{1A \cdot 1T}{1.6 \times 10^{-19} C \cdot 8.42 \times 10^{28} m^{-3} \cdot 25 \times 10^{-6} m} = 3.0 \times 10^{-6} V \quad \textbf{(Equation 1-11)}$$

Note the conversion of all quantities to SI (meter-kilogram-second) units for consistency in the calculation.

Even for the case of a magnetic field as strong as 10,000 gauss, the voltage resulting from the Hall effect is extremely small. For this reason, it is not usually practical to make Hall-effect transducers with most metals.

1.3 The Hall Effect in Semiconductors

From the previous description of the Hall effect in metals, it can be seen that one means of improvement might be to find materials that do not have as many carriers per unit volume as metals do. A material with a lower carrier density will exhibit the Hall effect more strongly for a given current and depth. Fortunately, semiconductor materials such as silicon, germanium, and gallium-arsenide provide the low carrier densities needed to realize practical transducer elements. In the case of semiconductors, carrier density is usually referred to as *carrier concentration*.

Table 1-1: Intrinsic carrier concentrations at 300°K [Soc185]

Material	Carrier Concentration (cm^{-3})
Copper (est.)	8.4×10^{22}
Silicon	1.4×10^{10}
Germanium	2.1×10^{12}
Gallium-Arsenide	1.1×10^7

As can be seen from Table 1-1, these semiconductor materials have carrier concentrations that are orders of magnitude lower than those found in metals. This is because in metals most atoms contribute a conduction electron, whereas the conduction electrons in semiconductors are more tightly held. Electrons in a semiconductor only become available for conduction when they acquire enough thermal energy to reach a conduction state; this makes the carrier concentration highly dependent on temperature.

Semiconductor materials, however, are rarely used in their pure form, but are doped with materials to deliberately raise the carrier concentration to a desired level. Adding a substance like phosphorous, which has five electrons in its outer orbital (and appears in column V of the periodic table) adds electrons as carriers. This results in what is known as an N-type semiconductor. Similarly, one can also add positive charge carriers by doping a semiconductor with column-III materials (three electrons in the outer orbital) such as boron. While this doesn't mean that there are free-floating protons available to carry charge, adding a column-III atom removes an electron from the semiconductor crystal to create a "hole" that moves around and behaves as if it were actually a charge-carrying particle. This type of semiconductor is called a P-type material.

For purposes of making Hall transducers, there are several advantages to using doped semiconductor materials. The first is that, because of the low intrinsic carrier concentrations of the pure semiconductors, unless materials can be obtained with part-per-trillion purity levels, the material will be doped anyhow—but it will be unknown with what or to what degree.

The second reason for doping the material is that it allows a choice of the predominant charge carrier. In metals, there is no choice; electrons are the default charge carriers. However, semiconductors there is the choice of either electrons or holes. Since electrons tend to move faster under a given set of conditions than holes, more sensitive Hall transducers can be made using an N-type material in which electrons are the majority carriers than with a P-type material in which current is carried by holes.

The third reason for using doped materials is that, for pure semiconductors, the carrier concentration is a strong function of temperature. The carrier concentration resulting from the addition of dopants is mostly a function of the dopant concentration, which isn't going to change over temperature. By using a high enough concentration of dopant, one can obtain relatively stable carrier concentrations over temperature. Since the Hall voltage is a function of carrier concentration, using highly doped materials results in a more temperature-stable transducer.

In the case of Hall transducers on integrated circuits, there is one more reason for using doped silicon—mainly because that's all that is available. The various silicon layers used in common IC processes are doped with varying levels of N and P materials, depending on their intended function. Layers of pure silicon are not usually available as part of standard IC fabrication processes.

1.4 A Silicon Hall-Effect Transducer

Consider a Hall transducer constructed from N-type silicon that has been doped to a level of 3×10^{15} cm^{-3}. The thickness is 25µm and the current is 1 mA. By substituting the relevant numbers into Equation 1-10, we can calculate the voltage output for a 1-tesla field:

$$V_H = \frac{0.001A \cdot 1T}{1.6 \times 10^{-19}C \cdot 3 \times 10^{21} \, m^{-3} \cdot 25 \times 10^{-6} \, m} = 0.083V \qquad \textbf{(Equation 1-12)}$$

The resultant voltage in this case is 83 mV, which is more than 20,000 times the signal of the copper transducer described previously. Equally significant is that the necessary bias current is 1/1000 that used to bias the copper transducer. Millivolt-level output signals and milliamp-level bias currents make for practical sensors.

While one can calculate transducer sensitivity as a function of geometry, doping levels, and bias current, there is one detail we have ignored to this point: the resistance of the transducer. While it is possible to get tremendous sensitivities from thinly doped semiconductor transducers for milliamps of bias current, it may also require hundreds of volts to force that current through the transducer. The resistance of the Hall transducer is a function of the conductivity and the geometry; for a rectangular slab, the resistance can be calculated by:

$$R = \frac{\sigma \cdot l}{w \cdot d} \qquad \textbf{(Equation 1-13)}$$

where
  R is resistance in ohms
  σ is the resistivity in ohm-cm
  l is the length in cm
  w is the width in cm
  d is the thickness in cm

In the case of metals, σ is a characteristic of the material. In the case of a semi-conductor, however, σ is a function of both the doping and a property called carrier mobility. Carrier mobility is a measure of how fast the charge carriers move in response to an electric field, and varies with respect to the type of semiconductor, the dopant concentration level, the carrier type (N or P type), and temperature.

In the case of the silicon Hall-effect transducer described above (d = 25 μm = 0.0025 cm), made from N-type silicon doped to a level of $3 \times 10^{15} cm^{-3}$, σ ≈ 1.7 Ω-cm at room temperature. Let us also assume that the transducer is 0.1 cm long and 0.05 cm wide. The resistance of this transducer is given by:

$$R = \frac{\sigma \cdot l}{w \cdot d} = \frac{1.7\,\Omega \cdot cm \times 0.1\,cm}{0.05\,cm \times 0.0025\,cm} = 1360\,\Omega \qquad \textbf{(Equation 1-14)}$$

With a resistance of 1360Ω, it will take 1.36V to force 1 mA of current through the device. This results in a power dissipation of 1.36 mW, a modest amount of power that can be easily obtained in many electronic systems. The sensitivity and power consumption offered by Hall-effect transducers made from silicon or other semiconductors makes them practical sensing devices.

Chapter 2

Practical Transducers

In the last chapter we examined the physics of a Hall-effect transducer and related its performance to physical characteristics and materials properties. While this level of detail is essential for someone designing a Hall-effect transducer, it is not necessary for someone attempting to design with an already existing and adequately characterized device.

2.1 Key Transducer Characteristics

What are the key characteristics of a Hall-effect transducer that should be considered by a sensor designer? For the vast majority of applications, the following characteristics describe a Hall-effect transducer's behavior to a degree that will allow one to design it into a larger system.

- Sensitivity
- Temperature coefficient (tempco) of sensitivity
- Ohmic offset
- Temperature coefficient of ohmic offset
- Linearity
- Input and output resistance
- Temperature coefficient of resistance
- Electrical output noise

Sensitivity

Transducer sensitivity, or gain, was the major focus of much of the last chapter, in which we analyzed the device physics. From a designer's standpoint, more sensitivity is usually a good thing, as it increases the amount of signal available to work with. A sensor that provides more output signal often require simplers and less expensive support electronics than one with a smaller output signal.

Because the sensitivity of a Hall-effect transducer is dependent on the amount of current used to bias it, the sensitivity of a device needs to be described in a way that takes this into account. Sensitivity can be characterized in two ways:

1. Volts per unit field, per unit of bias current (V/B × I)
2. Volts per unit field, per unit of bias voltage (1/B)

Since a Hall-effect transducer is almost always biased with a constant current, the first characterization method provides the most detailed information. Characterizing by bias voltage, however, is also useful in that it quickly tells you the maximum sensitivity that can be obtained from that transducer when it is used in a bias circuit operating from a given power-supply voltage.

Temperature Coefficient of Sensitivity

Although a Hall-effect transducer has a fairly constant sensitivity when operated from a constant current source, the sensitivity does vary slightly over temperature. While these variations are acceptable for some applications, they must be accounted for and corrected when a high degree of measurement stability is needed. Figure 2-1 shows the variation in sensitivity for an F.W. Bell BH-200 instrumentation-quality indium-arsenide Hall-effect transducer when biased with a constant current. The mean temperature coefficient of sensitivity of this device is about –0.08%/°C.

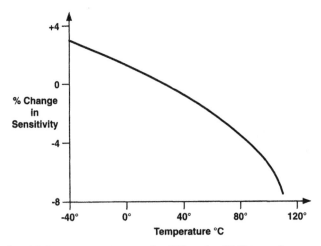

Figure 2-1: Sensitivity vs. temperature for BH-series Hall transducer under constant current bias (after [Bell]).

When operating a Hall-effect transducer from a constant-voltage bias source, one will obtain sensitivity variations over temperature considerably greater than those obtained when operating the device from a constant-current bias source. For this reason constant-current bias is normally used when one is concerned with the temperature stability of the sensor system. For the BH-200 device described above, constant-voltage bias would result in a temperature coefficient of sensitivity of approximately –0.2%/°C.

Ohmic Offset

Because we live in an imperfect world, we can't expect perfection in our transducers. When a Hall-effect transducer is biased, a small voltage will appear on the output even in the absence of a magnetic field. This offset voltage is undesirable, because it limits the ability of the transducer to discriminate small steady-state magnetic fields. A number of effects conspire to create this offset voltage. The first is alignment error of the sense contacts, where one is further "upstream" or "downstream" in the bias current than the other. Inhomogeneities in the material of the transducer can be another source. These effects are illustrated in Figure 2-2. Finally, the semiconductor materials used to make Hall-effect transducers are highly piezoresistive, meaning that the electrical resistance of the material changes in response to mechanical distortion. This causes most Hall-effect transducers to behave like strain gauges in response to mechanical stresses imposed on them by the packaging and mounting.

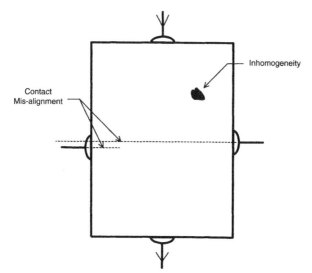

Figure 2-2: Ohmic offsets result from misalignment of the sense contacts and inhomogeneities in the material.

Although offset is usually expressed in terms of output voltage for a given set of bias conditions, it also needs to be considered in terms of magnetic field units. For example, compare a transducer with 500 μV of offset and a sensitivity of 100 μV/gauss, to a second transducer with 200 μV of offset but only 10 μV/gauss of sensitivity. The first transducer has a 5-gauss offset while the latter has an offset of 20 gauss, in addition to having a much lower sensitivity. For applications where low magnetic field levels are to be measured, the first sensor would tend to be easier to use, both because it provides a higher sensitivity and also because it provides a lower offset error when considered in terms of the quantity being measured, namely magnetic field.

Temperature Coefficient of Ohmic Offset

Like sensitivity, the offset of a transducer will drift over temperature. Unlike sensitivity, however, the offset drift will tend to be random, varying from device to device, and is not generally predictable. Some offset drift results from piezoresistive effects in the transducer. As temperature varies, uneven expansion of the materials used to fabricate a transducer will induce mechanical stresses in the device. These stresses are then sensed by the Hall-effect transducer. In general, devices with larger initial offsets also tend to have higher levels of offset drifts. While there are techniques for minimizing offset and its drift, precision applications often require that each transducer be individually characterized over a set of environmental conditions and a compensation scheme be set for that particular transducer.

Linearity

Because Hall-effect transducers are fundamentally passive devices, much like strain gauges, the output voltage cannot exceed the input voltage. This results in a roll-off of sensitivity as the output voltage approaches even a small fraction of the bias voltage. In cases where the Hall voltage is small in comparison to the transducer bias voltage, Hall sensors tend to be very linear, with linearity errors of less than 1% over significant operating ranges. When constructing instrumentation-grade sensors, which are expected to measure very large fields such as 10,000 or even 100,000 gauss, it is often desirable to use low sensitivity devices that do not easily saturate.

Input and Output Resistances

These parameters are of special interest to the circuit designer, as they influence the design of the bias circuitry and the front-end amplifier used to recover the transducer signal. The input resistance affects the design of the bias circuitry, while the output resistance affects the design of the amplifier used to detect the Hall voltage. Although it is possible to design a front-end amplifier with limited knowledge of the resistance of the signal source that will be feeding it, it may be far from optimal from either performance or cost standpoints, compared to an amplifier designed in light of this information.

For low-noise applications, the output resistance is of special interest, since one source of noise, to be discussed later, is dependent on the output resistance of the device. A simple electrical model describing a Hall-effect transducer from a circuit-interface standpoint is presented later on in Chapter 3.

Temperature Coefficient of Resistance

The temperature coefficient of the input and output resistances will either be identical, or match very closely. Knowing the temperature variation of the input resistance is useful when designing the current source used to bias the transducer. For transducers biased with a constant current source, the bias voltage will be proportional to the transducer's resistance. A bias circuit designed to drive a transducer with a particular resistance at room temperature may fail to do so at hot or cold extremes if variations in transducer resistance are not anticipated. For practical transducers the temperature coefficient of resistance can be quite high, often as much as 0.3%/°C. Over an automotive temperature range (−50° to +125°C) this means that the input and output resistances can vary by as much as 30% from their room temperature values.

Noise

In addition to providing a signal voltage, Hall-effect transducers also present electrical noise at their outputs. For now we will limit our discussion to sources of noise actually generated by the transducer itself, and not those picked up from the outside world or developed in the amplifier electronics.

The most fundamental and unavoidable of electrical noise sources is called Johnson noise, and it is the result of the thermally induced motion of electrons (or other charge carriers) in a conductive material. It is solely a function of the resistance of the device and the operating temperature. Johnson noise is generated by any resistance (including that found in a Hall transducer), and is described by:

$$V_n = \sqrt{4kTRB}$$
(**Equation 2-1**)

where k is Boltzmann's constant (1.38×10^{-23} K^{-1})
T is absolute temperature in °K
R is resistance in ohms
B is bandwidth in hertz

The bandwidth over which the signal is examined is an important factor in how much noise is seen. The wider the frequency range over which the signal is examined, the more noise will be seen.

The down side of Johnson noise is that it defines the rock-bottom limit of how small a signal can be recovered from the transducer. The two positive aspects are that it can be minimized by choice of transducer impedance, and that it is not usually of tremendous magnitude. A 1-kΩ resistor, for example, at room temperature (300°K) will only generate about 400 nanovolts RMS (root-mean-squared) of Johnson noise measured across a 10-kilohertz bandwidth.

Flicker noise, also known as $1/f$ noise, is often a more significant problem than Johnson noise. This type of noise is found in many physical systems, and can be generated by many different and unrelated types of mechanisms. The common factor, however, is the resultant spectrum. The amount of noise per unit of bandwidth is, to a first approximation, inversely proportional to the frequency; this is why it's also referred to as $1/f$ noise. Because many sensor applications detect DC or near-DC low-frequency signals, this type of noise can be especially troublesome. Unlike Johnson noise, which is intrinsic to any resistance regardless of how it was constructed, the flicker noise developed by a transducer is related to the specific materials and fabrication techniques used. It is therefore possible to minimize it by improved materials and processes.

The following sections describe the construction and characteristics of several types of Hall-effect transducers that are presently in common use.

2.2 Bulk Transducers

A bulk-type transducer is essentially a slab of semiconductor material with connections to provide bias and sense leads to the device. The transducer is cut and ground to the desired size and shape and the wires are attached by soldering or welding. One advantage of bulk-type devices is that one has a great deal of choice in selecting materials. Another advantage is that the large sizes of bulk transducers result in lower impedance levels and consequently lower noise levels than those offered by many other processes. Some key characteristics of an instrument-grade bulk indium-arsenide transducer (the F.W. Bell BH-200) are shown in Table 2-1.

Table 2-1: Key characteristics of BH-200 Hall Transducer

Characteristic	Value	Units
Nominal bias current	150	mA
Sensitivity at recommended bias current (Ibias = 150 mA)[1]	15	µV/G
Sensitivity (current referenced)[1]	100	µV/G•A
Temperature coefficient of sensitivity	−0.08	%/°C
Ohmic offset, electrical (maximum) – (Ibias = 150 mA)	±100	µV
Ohmic offset, magnetic (maximum)[1]	±7	gauss
Tempco of ohmic offset, electrical – (Ibias = 150 mA)	±1	µV/°C

(Continued)

Characteristic	Value	Units
Tempco of ohmic offset, magnetic[1]	±0.07	±gauss/°C
Max linearity error (over ±10 kilogauss)	±1	%
Input resistance (max)	2.5	Ω
Output resistance (max)	2	Ω
Temperature coefficient of resistance	0.15	%/°C

Note 1: These parameters estimated from manufacturer's data

2.3 Thin-Film Transducers

A thin-film transducer is constructed by depositing thin layers of metal and semiconductor materials on an insulating support structure, typically alumina (Al_2O_3) or some other ceramic material. Figure 2-3 provides an idealized structural view of a "typical" thin-film Hall-effect transducer. The thickness of the films used to fabricate these devices can be on the order of 1 μm or smaller.

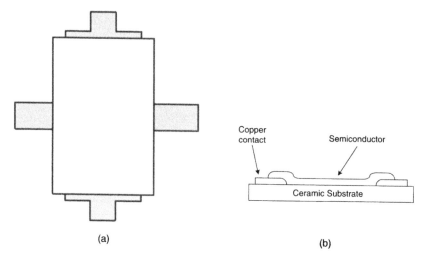

(a)   (b)

Figure 2-3: Schematic top (a) and cross-section (b) views of thin-film Hall-effect transducer.

The primary advantages of thin-film construction are:
- Flexibility in material selection
- Small transducer sizes achievable
- Thin Hall-effect transducers provide more signal for less bias current
- Photolithographic processing allows for mass production

Each layer is added to the thin-film device by a process that consists of covering the device with the film and selectively removing the sections that are not wanted, leaving the desired patterns. The details of the processing operations for each layer vary depending on the characteristics of the materials being used.

Film deposition is commonly accomplished through a number of means, the two most common being evaporation and sputtering. In evaporation, the substrate to be coated and a sample of the coating material are both placed in a vacuum chamber, as shown in Figure 2-4. The sample is then heated to the point where it begins to vaporize into the vacuum. The vapor then condenses on any cooler objects in the chamber, such as the substrate to be coated. Because the hot vapor in many cases will chemically react with any stray gas molecules, a substantially good vacuum is required to implement this technique. Vacuums of 10^{-6} to 10^{-7} torr (760 torr = 1 atmosphere) are commonly required for this type of process. The thickness of the deposited film is controlled by the exposure time.

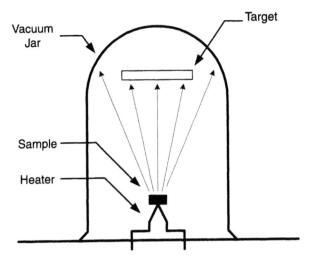

Figure 2-4: Schematic drawing of evaporative thin-film process.

Sputtering is another method for coating substrates with thin films. In sputtering, the sample coating material is not directly heated. Instead, an inert gas, such as argon, is ionized into a plasma by an electrical source. The velocity of the ions of the plasma is sufficient to knock atoms out of the coating sample (the target), at which point they can deposit themselves on the substrate to be coated. As in the case of an evaporative coating system, film thickness is controlled by exposure time. Figure 2-5 shows a schematic view of a sputtering system.

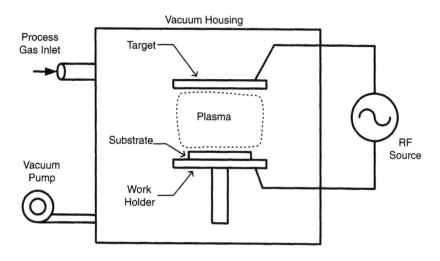

Figure 2-5: Sputtering method of thin-film deposition.

The principal advantage offered by sputtering over evaporation is that, because the coating material doesn't need to be heated to near its evaporation point, it is possible to make thin films with a much wider variety of materials than is possible by evaporation.

Once a material has been laid down in a thin film, it is then patterned with a photoresist material, exposed to a photographic plate carrying the desired pattern, and the photoresist is then developed, leaving areas of the substrate selectively exposed. The substrate is then etched, often by immersion in a suitable liquid solvent or acid. Alternatively, the substrate can be plasma-etched by a process related to sputtering. In either case, after the etching step is finished, the remaining photoresist is stripped and the substrate is prepared to receive the next layer of film or readied for final processing. The sequence of operations needed to process a layer of a thin film is summarized in Figure 2-6.

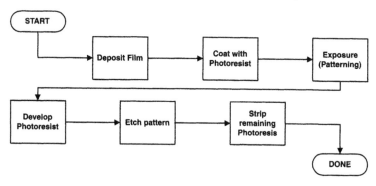

Figure 2-6: Thin-film processing sequence.

The HS-100 is an example of a commercial thin-film Hall transducer manufactured by F.W.Bell. This device is made with two thin-film layers, a metal layer to provide contacts to the Hall-effect element, and an indium-arsenide thin film that forms the Hall-effect transducer itself. In addition, solder bumps are deposited on the copper to provide connection points to the outside world. Wires may be soldered to these features, or the device may be placed face down on a printed circuit board or ceramic hybrid circuit, and reflow soldered into place.

Key specifications for the HS-100 transducer are listed in Table 2-2. The major improvements over bulk devices are in the area of sensitivity and supply current; the thin-film device is nearly as sensitive as the previously described bulk device (BH-200), and obtains this level of sensitivity with an order of magnitude less supply current needed. The BH-200 bulk device, however, is superior in the areas of offset error and drift over temperature. The principal advantage of the thin-film device is potentially lower cost. Thin-film processing techniques allow a great number of devices to be fabricated simultaneously and separated into individual units at the end of processing.

Table 2-2: Key characteristics of HS-100 Hall transducer

Characteristic	Value	Units
Nominal bias current	150	mA
Sensitivity at recommended bias current (Ibias = 10 mA)[1]	8	μV/gauss
Sensitivity (current referenced)[1]	800	μV/gauss•A
Temperature coefficient of sensitivity	–0.1	%/°C
Ohmic offset, electrical (maximum) (Ibias = 10 mA)	±6	mV
Ohmic offset, magnetic (maximum)[1]	±750	Gauss
Tempco of ohmic offset, electrical (Ibias = 10 mA)	±10	μV/°C
Tempco of ohmic offset, magnetic[1]	±1.25	±gauss/°C
Max linearity error (over ±10 kilogauss)	—	%
Input resistance (max)	160	Ω
Output resistance (max)	360	Ω
Temperature coefficient of resistance	0.1	%/°C

Note 1: These parameters estimated from manufacturer's data

2.4 Integrated Hall Transducers

Making a Hall-effect transducer out of silicon, using standard integrated circuit processing techniques, allows one to build complete sensor systems on a chip. The transducer bias circuit, the front-end amplifier, and in many cases application-specific signal processing can be combined in a single low-cost unit. The addition of electronics to the bare transducer allows sensor manufacturers to provide a very high degree of functionality and value to the end user, for a modest price. By simultaneously fabricating thousands of identical devices on a single wafer, it is possible to economically produce large numbers of high-quality sensors. Figure 2-7 shows an example of several hundred Hall-effect sensors on a silicon wafer, before being separated and individually packaged.

Figure 2-7: Hall-effect sensor ICs on a silicon wafer. *(Courtesy of Melexis USA.)*

While there are many layers and structures available in modern integrated circuit processes that can be exploited to fabricate Hall-effect transducers, we will illustrate the basics by considering one particular case. Because of the complexity of integrated circuit fabrication processes, we will not even attempt to describe them here. Interested readers will find good descriptions of how silicon ICs are made in [GRAY84]. For this example, we will consider what is known as an epitaxial Hall-effect transducer. We will begin by considering the structure of a related device, the epitaxial resistor.

Figure 2-8 shows top views and side views of an epitaxial resistor made with a typical bipolar process. The device is so named because it is built in the epitaxial N-type silicon layer. The raw wafer is usually of a P-type material, and the epitaxial layer is deposited on the surface of the wafer by a chemical vapor deposition (CVD) process, and can be doped independently of the raw wafer. P-type isolation walls are then implanted or diffused into the top surface of the epitaxial layer to form wells (isolated islands) of N-type material. Maintaining each of the wells at a positive voltage with respect to the P-type substrate causes the P-N junctions to be reverse-biased, thus electrically isolating the wells from each other. By providing this junction isolation, one can build independent circuit components such as resistors, transistors, and Hall-effect transducers, in a single, monolithic piece of silicon, using the wells as starting points. The overall depth of epitaxial layers can vary from 2–30 μm for commonly available IC processes.

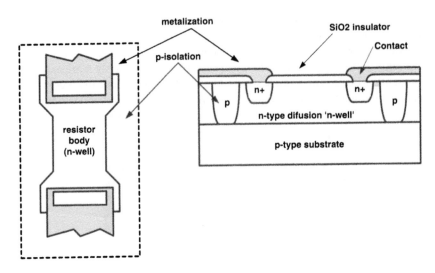

Figure 2-8: Structure of epitaxial resistor showing various layers.

In the case of an epitaxial resistor, the well defines the body of the component. The N-type material used typically has a resistance of about 2–5 kΩ when measured across the opposite edges of a square section. This allows one to readily construct resistors with values up to about 100 kΩ by building long, narrow resistor structures. The whole IC is then covered with an insulating layer of SiO_2 (silica glass), and holes called contact windows are then etched through this glass layer at specific points to allow for electrical contact to the underlying silicon. Finally a layer of aluminum is patterned on top of the SiO_2 to make to form the "wiring" for the IC, with the metal extending down through the contact windows to connect to the silicon. To get a good electrical contact with the aluminum, a plug of high concentration N-type material (somewhat confusingly referred to as "N+" material) is driven into the epitaxial resistor just under the contact areas before the SiO_2 is grown over the device.

"Epi" resistors, as they are commonly called, are easy to make in a bipolar process because they require no additional process steps beyond those required to make NPN transistors. Their performance characteristics, however, are fairly awful, at least when compared to the discrete resistors most electronic designers commonly use. Their absolute tolerance is on the order of ±30%, and they experience temperature coefficients of up to 0.3%/°C. In addition, the effective thickness of the reverse-biased P-N junction that isolates an epi resistor from the substrate varies with applied voltage. This has the effect of making the resistor's value dependent on the voltage applied at its terminals.

Despite the drawbacks of using the epitaxial layer when making resistors, it is quite useful for making Hall-effect transducers. Because the epitaxial layer is relatively thin (5 μm is thin from a macroscopic perspective), and usually made from lightly ($N = 10^{15}/cm^3$) doped silicon, it is possible to make reasonably sensitive Hall-effect transducers that have modest power requirements.

Because IC manufacturers view the exact details of their processes as trade secrets and are thus not inclined to broadcast them to the world, we will present an example of a Hall device fabricated with a "generic" bipolar process. This will give a general idea of the performance one can expect from such a device. Figure 2-9 shows the details of this device.

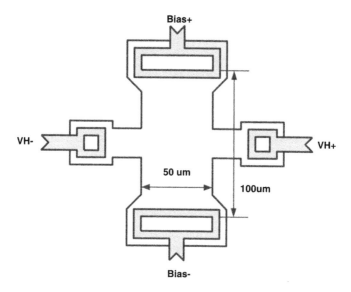

Figure 2-9: Integrated epitaxial Hall-effect transducer layout.

Note that the connections of the sense terminals are made by bringing out "ears" of epitaxial material from the body of the device and making metal contact at these points, instead of simply placing the sense contacts directly on the transducer. There are two reasons for doing this: the first is to maximize the sensitivity by ensuring that all the bias current flows between the sense terminals, and the second is to minimize ohmic offsets. Because the "ears" are fabricated in the same process step as the rest of the transducer, a high degree of alignment is naturally maintained. Metalization and contact windows, on the other hand, are fabricated in separate manufacturing steps, increasing the opportunities for contact misalignment.

For this transducer, the critical physical parameters are:
- Length = 200 μm
- Width = 100 μm
- Thickness (of epi layer) = 10 μm

- Carrier concentration = $3 \times 10^{15}/cm^3$ ($3 \times 10^{21}/m^3$)
- Bulk resistivity $\sigma = 2$ Ω-cm (0.02 Ω-m)

The sensitivity per unit of current and field can be calculated by using equation 1-10, yielding:

$$V_H = \frac{IB}{q_0 Nd} = \frac{1A \cdot 1T}{1.6 \times 10^{-19} C \cdot 3 \times 10^{21} m^{-3} \cdot 10^{-5} m} = 208V \qquad \textbf{(Equation 2-1)}$$

for 1 tesla at 1 ampere of bias, or 20.8 mV/G-A in cgs units. This is an amazingly high level of sensitivity. This sensor, however, will never operate at one ampere; 1 milliampere is a more realistic bias current. Even at 1 milliampere, however, this transducer will still provide 20 μV/gauss.

The next major question is that of input and output resistance. Because the bias current flows in a substantially uniform manner between the bias contacts, since the contacts extend across the width of the transducer, we can make a fairly good estimate of the input resistance by:

$$R_{in} = \sigma \frac{L}{W \cdot T} = 0.02\Omega \cdot m \frac{200 \times 10^{-6} m}{100 \times 10^{-6} m \cdot 10^{-5} m} = 4000\Omega \qquad \textbf{(Equation 2-2)}$$

It therefore requires 4V to bias the transducer with 1 mA.

Because of geometric factors, the output resistance cannot be as readily calculated as the input resistance. If one were to apply a voltage between the output terminals, the lines of current flow would not be parallel and uniform (and therefore amenable to back-of-envelope analysis). For purposes of designing a compatible front-end amplifier and noise calculations, one might assume that the resistance of the output is within a factor of two or three of that of the input.

Because the estimation of temperature sensitivities and ohmic offset is very difficult (if not impossible) even when one is working with a fully characterized process, we shall ignore them. Suffice it to say, however, that integrated Hall-effect transducers can be made with substantially good performance in these areas. For sake of comparison with the previous examples, Table 2-3 lists a few of the predicted and "guesstimated" characteristics of our hypothetical integrated transducer.

Table 2-3: Key characteristics of hypothetical silicon integrated Hall-effect transducer

Characteristic	Value	Units
Sensitivity at recommended bias current (Ibias = 1 mA)	20	μV/G
Sensitivity (current referenced)	20	mV/G•A

(Continued)

Characteristic	Value	Units
Temperature coefficient of sensitivity (for constant-current bias)	−0.1	%/°C
Ohmic offset, electrical (maximum) (Ibias = 1 mA)	±10	mV
Ohmic offset, magnetic (maximum)	±500	G
Max linearity error (over ±1 kilogauss)	1	%
Input resistance	4000	Ω
Output resistance	4000?	Ω
Temperature coefficient of resistance	0.3	%/°C

It is also possible to construct integrated Hall-effect transducers from gallium-arsenide, germanium, and other semiconductor materials for even better performance. Integrated processes based on these other materials, however, do not provide the wealth of electronic device types that can be cofabricated on silicon processes.

Silicon processes have another advantage: availability. High-quality Hall-effect transducers can be fabricated with many standard bipolar and CMOS integrated circuit processes with little or no modification. A number of semiconductor companies presently produce a vast array of Hall-effect integrated circuits.

Figure 2-10 shows an example of a silicon Hall-effect IC, containing a Hall-effect transducer and a number of other components such as transistors and resistors. The Hall-effect transducer is the square-shaped object in the center. The size of this IC is roughly 1.5 mm × 2 mm.

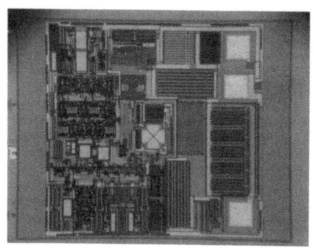

Figure 2-10: Silicon Hall-effect sensor IC with supporting electronics. *(Courtesy of Melexis USA.)*

2.5   Transducer Geometry

To this point, we have largely ignored the role of geometry in the construction of a Hall-effect transducer. The specific geometry used which device fabrication, however, can have a large impact on its performance and consequent suitability as a component.

The main factors that can be optimized by transducer geometry are sensitivity, offset, and power consumption. Let us examine the rectangular slab form as a starting point for improvements.

In the rectangular transducer form (Figure 2-11a), a uniform current sheet is established by bias electrodes that run the width of the device. Since the sensitivity is proportional to the total current passing between the sense electrodes, it would at first glance seem that by either making the sensor wider or shorter, more bias current could be driven through the device for a given bias voltage. More bias current does flow in these cases, but the wide bias electrodes form a low-resistance path to short-circuit the Hall voltage. For similar reasons, chaining multiple Hall transducers so that the bias terminals are connected in parallel and the output terminals are in series does not significantly increase the output sensitivity. For a rectangular transducer, maximum sensitivity for a given amount of power dissipation is achieved when the ratio of length to width is about 1.35 [Baltes94].

One method of avoiding end-terminal shorting is to use a cross pattern (Figure 2-11b). Because the input resistance rises rapidly with the lengthening of the cross, this geometry is not a particularly good one to use when trying to optimize sensitivity.

Another method of reducing end-terminal shorting is through the use of a diamond-shaped transducer (Figure 2-11c). In this device, all the terminals are essentially points, and the current spreads though the device in a nonuniform manner. Although the diamond shape is not optimal from a sensitivity standpoint, it offers other advantages; one of the major advantages is that the sense terminals out at the edges of of the current bias the transducer. In this respect the diamond shape works well; because the current flow at the sense corners of the diamond is low, the voltage gradient in the corners will also be low. This tends to reduce ohmic offset from contact misalignment effects.

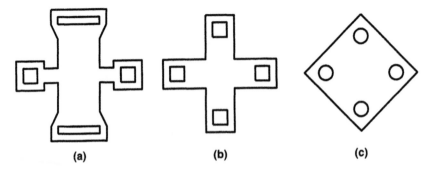

(a)   **(b)**   **(c)**

Figure 2-11: Common Hall transducer shapes: rectangle (a), cross (b), diamond (c),

2.6 The Quad Cell

In integrated Hall-effect transducers, where features can be defined with very high (submicron) resolutions, geometric flaws can be a minor source of output offset voltage. Three additional and significant sources of offset are:

- Process variation over the device
- Temperature gradients across the device in operation
- Mechanical stress imposed by packaging.

Process variations such as the amount and depth of doping can vary slightly over the surface of a wafer, leading to very slight nonuniformities between individual devices. In the case of some components, such as resistors, this effect is most readily seen as a degree of mismatch between two proximate and identical devices. In a device such as a Hall-effect transducer, this effect manifests itself as offset voltage errors. If the transducer is thought of as a balanced resistive bridge, as shown in Figure 2-12, inconsistencies appear as ΔR in one or more of the legs.

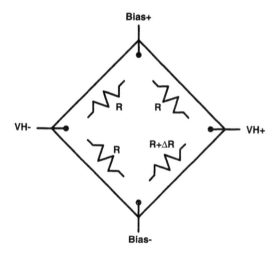

Figure 2-12: Transducer offset errors modeled as imbalanced resistive bridge.

When an integrated circuit is operating, the power dissipated in the device causes heating of the silicon die. Because most circuits dissipate more power in some parts than in others, the heating is not uniform. The resultant temperature differences can cause identical devices to behave differently, depending on where they are situated and their actual operating temperature. In some cases, in addition to being sensitive to their absolute temperature, a device may exhibit different behavior in response to temperature gradients appearing across it. While it may be difficult to believe that temperature gradients across a microscopic structure can be significant, consider that a matched pair of devices with temperature coefficients of resistance on the order of 0.3%/°C only need differ in temperature by about 1/3°C to create a 0.1% mismatch.

Finally, silicon is a highly piezoresistive material, meaning its resistance changes when you mechanically deform it. While this effect is useful when making strain gauges, it is a nuisance when making magnetic sensors. Mechanical stresses in an IC come from a number of sources, but primarily result from the packaging. The silicon die, the metal leadframe, and the plastic housing all have slightly different thermal coefficients of expansion. As the temperature of the packaged IC is varied, this can result in enormous compressive and shear stresses being applied to the surface of the IC chip. In extreme cases this can actually result in damaging the IC chip, even to the extent of fracturing it. Additionally, the processes used for molding "plastic" packages around ICs tend to leave considerable residual stresses in the package after the overmolding material cools and sets.

The technique of "Quadding" [Bate79] offers substantial immunity to offset effects from the above three sources of offset. While these effects behave in very complicated ways, if one assumes that they behave either uniformly or linearly over very small regions of an IC, such as the transducer, one can use the offsets induced in one device to cancel out those induced in an adjacent device.

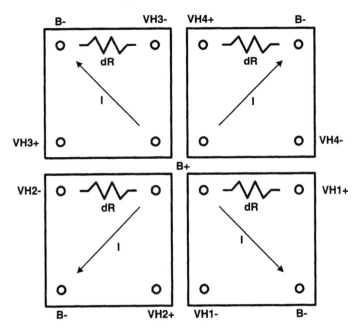

Figure 2-13: "Quad" transducer layout causes identical offsets to cancel each other out.

Figure 2-13 shows a Hall-effect transducer using a quadded layout. If one assumes that the effect causing the offset will create the offset equally in the four separate transducers, then the ΔR will occur in the same physical leg of each device, and will result in a ΔV in addition to the Hall voltage from that device. The individual voltages seen at the outputs of the individual devices will be:

$$V_1 = V_H + \Delta V$$
$$V_2 = V_H - \Delta V$$
$$V_3 = V_H + \Delta V$$
$$V_4 = V_H - \Delta V \qquad \text{(Equation 2-3)}$$

The transducers are then wired so that their signals are averaged. This results in an output signal of just V_H, with no offset error, at least in theory. In practice you still will get some offset voltage with a quadded transducer, but it will be an order of magnitude smaller than that obtained from a single device. Figure 2-14 shows how the devices can be wired in parallel. Similar wiring schemes have often been used because none of the wires cross, meaning that the transducer can be readily implemented in IC processes that only provide a single layer of metal for component interconnection.

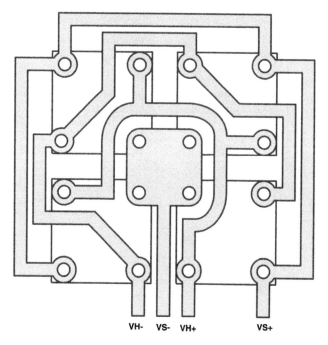

Figure 2-14: Parallel wiring structure for quadded Hall transducer.

2.7 Variations on the Basic Hall-Effect Transducer

Although most commercial Hall-effect devices employ the types of transducers previously described in this chapter, several variations on the basic technology have been developed that offer additional performance and capabilities. The two most significant of these technologies are the vertical Hall transducer and the incorporation of integrated flux concentrators.

One of the fundamental limitations of traditional Hall-effect transducers is that they provide sensitivity in only one axis—the one perpendicular to the surface of the IC on which they are fabricated. This means that to sense field components in more than one axis, one needs to use more than one sensor IC, and those sensor ICs must be individually mounted and aligned. For example, in order to realize a three-axis magnetic sensor with traditional Hall-effect transducers, three separate devices must be used, and the designer must try to align them along the desired sensing axis, all while trying to maintain close physical proximity. While this is not impossible, it can be difficult and expensive to implement such a sensor, especially if the transducers need to be physically close together.

The vertical Hall-effect transducer [Baltes94] is one means of providing multi-axis sensing capability on a single silicon die. Figure 2-15 shows the basic structure of this device.

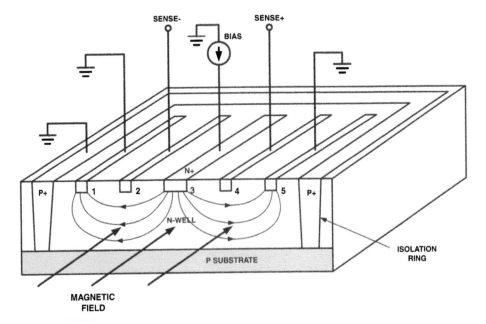

Figure 2-15: Vertical Hall-effect transducer (after Baltes et al.).

In the vertical Hall-effect transducer, bias current is injected into an N-well from a central terminal 3, and is symmetrically collected by ground terminals 1 and 5. The current path goes down from the central terminal, and arches across the IC and back up to the ground terminals. In the absence of an applied magnetic field, this current distribution results in equal potentials being developed at sense terminals 2 and 4.

When a magnetic field is applied across the face of the chip perpendicular to the current paths, Lorentz forces cause a slight shift in the current paths, as they do in a traditional Hall-effect transducer. This in turn causes a voltage differential to be developed across the sense terminals, which can then be amplified and subsequently processed into a usable signal level.

Because the vertical Hall-effect transducer, like its more traditional cousin, is sensitive to field in a single axis, it is possible to fabricate a two-axis sensor by placing a pair of these devices on a single silicon die by aligning their structures at 90° rotation to each other. Finally, one can also add a conventional Hall-effect transducer to the same die to obtain a third axis of sensitivity. In this way, it becomes possible to create a three-axis magnetic transducer on a single silicon die.

One disadvantage of the vertical Hall structure is that it lacks four-way symmetry. As will be seen in the next chapter, transducer symmetry can be exploited at the system level to reduce the effects of ohmic offset voltage errors.

Another structure that offers significant advantages is the Hall-effect transducer with *integrated magnetic flux concentrators* (IMCs) [Popovic01]. A magnetic flux concentrator is a piece of ferrous material, such as steel, that is used to direct or intensify magnetic flux towards a sensing element. External flux concentrators have long been used externally to direct and concentrate magnetic flux in Hall-effect applications. The novel aspect of the IMC is in fabricating the flux concentrator on the surface of the silicon die in extremely close proximity to the Hall-effect transducer, as shown in Figure 2-16.

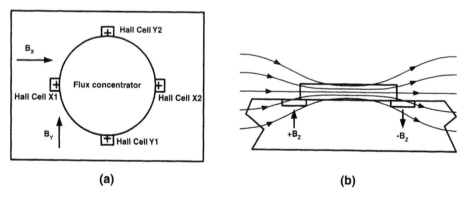

(a)   (b)

Figure 2-16: Integrated magnetic flux concentrator, top view (a) and side view (b).

The flux concentrator shown in Figure 2-16 would normally be implemented as a thin layer of a high-permeability magnetic alloy such as permalloy (a nickel-iron steel), which would be laid down on the IC surface with an evaporation or sputtering process. In the configuration shown, there are four Hall-effect transducers arranged around the periphery of the flux concentrator. The concentrator performs two functions. The first is to concentrate the field in its proximity (Figure 2-16b). This intensifies the field seen by the transducers and has the effect of increasing the transducers' effective sensitivity. The second function performed by the concentrator is to redirect the axis of the applied field from horizontal to vertical near the transducers. For example, a horizontally applied X field is mapped into a positive Z component at transducer X1 and a negative component at transducer X2. Note, however, that the transducers will still be sensitive to fields applied in the Z-axis despite the presence of the flux concentrator. By subtracting the outputs of the transducers (X2–X1, Y2–Y1), the effects of any Z-field components can be ignored. A microphotograph of an IMC Hall-effect transducer can be seen in Figure 2-17.

Figure 2-17: Microphotograph of IMC Hall-effect transducer. *(Courtesy Melexis USA.)*

Commercially available products utilizing IMC Hall-effect transducers have been developed by Sentron AG. Two typical devices are the CSA-1V-SO and the 2SA-10. The CSA-1V-SO is a single-axis device in an SOIC-8 package, while the 2SA-10 is a two-axis device. Both of these devices incorporate on-chip amplifier circuitry in addition to the transducer elements. The CSA-1V-SO provides a very high level of sensitivity, typically 30-mV output per gauss of applied field, and can sense fields over a range of approximately ±75 mV. Because of the device's high sensitivity and a sensing axis parallel to the SOIC package face, this device has potential for replacing magneto-resistive sensors in many applications. The 2SA-10 also is provided in an SOIC-8 package, and provides somewhat lower sensitivity (5-mV output per gauss of applied field), but also offers two sensing axes, both parallel to the SOIC face. The primary application for this device is in sensing rotary position, where one simultaneously measures field strength in two axes, and resolves the two measurements into degrees of rotation.

2.8 Examples of Hall Effect Transducers

Table 2-4 lists a few examples of commercially available Hall-effect transducers. Keep in mind that these devices are intended for a variety of applications, so sensitivity alone should not be used to determine a particular transducer's suitability for a particular use.

Table 2-4: Examples of commercial Hall-effect transducers

Manufacturer	Device	Sensitivity	Material
F.W. Bell	HS-100	8μV/G @ 10 mA	Thin-Film Indium Arsenide
	BH-200	15μV/G @ 150 mA	Bulk Indium-Arsenide
Asahi-Kasei	HG-106C	100μV/G @ 6 mA	Gallium Arsenide
Sprague Electric[1]	UGN-3604	60μV/G @ 3 mA	Silicon (monolithic)

Note 1: This product was discontinued a long time ago, and is only mentioned here to provide an example of a silicon Hall-effect transducer's "typical" sensitivity..

Chapter 3

Transducer Interfacing

While it is possible to use a Hall-effect transducer as a magnetic measuring instrument with merely the addition of a stable power supply and a sensitive voltmeter, this is not a typical mode of application. More frequently the transducer is used in conjunction with electronics specially designed to properly bias it and perform some preprocessing of the resultant signals before presenting them to the end-user. The addition of application-specific electronics provides significant value by allowing the end-user to view the system as a black-box, without having to concern himself with the details of how the transducer is implemented. To differentiate a bare transducer from a transducer with support electronics, we will be referring to the latter as a sensor.

A minimal Hall-effect sensor (Figure 3-1) consists of three parts: a means of powering or biasing the transducer, the transducer itself, and an amplification stage. Because of the variety of applications in which Hall-effect sensors are employed, and their equally diverse functional requirements, there is no single "best way" to build even a minimal transducer interface. The "goodness" of any implementation is a function of how well it meets the requirements of a particular application. These requirements can include sensing accuracy, cost, packaging, power consumption, response time, and environmental compatibility. A $4,000 laboratory gaussmeter would not be a good (or even adequate) solution under the hood of a car, nor would a 20-cent commodity sensor IC be an especially good choice for many laboratory applications; each has its own application domain for which it is best suited.

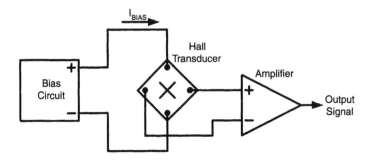

Figure 3-1: Minimal components of Hall-effect sensor system.

3.1 An Electrical Transducer Model

To design good interface circuitry for a transducer requires that one understand how the transducer behaves. While the first two chapters described the physics and construction of a number of Hall-effect transducers, there still remains the question of how it behaves as a circuit element. Carrier concentration, current density, and geometry describe the device from a physical standpoint, but what is needed is a model that describes how the device interacts with transistors, resistors, op-amps, and other components dear to the hearts of analog designers.

When confronted with an exotic component, such as a transducer, a good circuit designer will attempt to build a model to approximately describe that device's behavior, as seen by the circuits it will be connected to. For this reason, the model will usually be built from primitive electronic elements, and be represented in a highly symbolic (schematic) manner. The elements employed can include resistors, capacitors, inductors, voltage sources and current sources. There are several advantages inherent in this approach:

1) Circuit designers think in terms of electronic components and a good circuit-level model can allow a designer to understand the system. A great model can give a circuit designer the gut-level intuitive understanding of the system needed to produce first-rate work. Alternatively, a poor model can give a circuit designer gut-level feelings best resolved with antacids.

2) Simple circuit-level models often are analytically tractable. Deriving a set of closed-form analytic relationships can allow one to deliberately design to meet a set of goals and constraints, as opposed to designing through an iterative, generate-and-test procedure.

3) Circuits can be automatically analyzed on a computer, by a number of commercially available circuit simulation programs (e.g., SPICE). In the hands of a skilled designer, the use of these tools can result in robust and effective designs. Conversely, in the hands of unskilled designers, their use can result in mediocre designs reached by trial-and-error (also known as the *design by brute-force and ignorance* method).

Figure 3-2 shows the model that was initially presented in the last chapter. It consists of four resistors and two controlled voltage sources. This model describes the transducer's input and output resistances, as well as its sensitivity as a function of bias voltage.

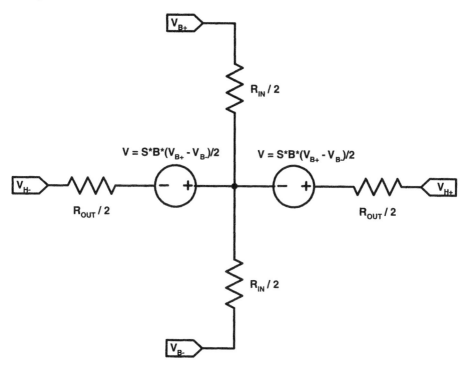

Figure 3-2: Hall-effect transducer simple electrical model.

The four resistors describe the input and output resistances of the transducer. In the case of a transducer with four-way symmetry, all of the resistors are equal. The voltage sources model the transducer's sensitivity or the gain, which is a linear function of the bias voltage and the applied magnetic field, The various variables and constants in this model are defined as follows:

V_{B+}, V_{B-}	Bias voltage
S	Sensitivity in $Vo/B * Vi$
B	Magnetic flux density
R_{IN}	Input resistance
R_{OUT}	Output resistance

Although this model is a gross simplification, it will exhibit enough of the electrical attributes of a transducer to be useful as an aid to designing interface circuits. The following are some of the major assumptions and limitations:

- Magnetic linearity; there are no saturation effects at high field.
- Temperature coefficients are ignored.
- There is no zero-flux offset.
- The resistance as measured between adjacent terminals is unimportant for many applications; modeling this correctly would unnecessarily complicate the model.
- A real Hall-effect sensor is a passive device; this model contains power-producing elements. We assume this additional power is small enough to ignore.
- The transducer is symmetric; the sense terminals are placed at the halfway point along the device.

3.2 A Model for Computer Simulation

The model presented in the last section can be adjusted so that it is suitable for simulation by SPICE (simulation program with integrated circuit emphasis) or another circuit simulation program. A few additional details need to be added both to make it more specific and to make it fit into SPICE's view of the world. Since SPICE doesn't directly handle magnetic field quantities, magnetic flux is represented by a voltage input to the model. SPICE also requires the user to define circuit topology by numbering each electrical node in the circuit. If you are using a graphical schematic-capture program to input your circuits, the computer numbers the nodes automatically. This can be a major convenience, especially when simulating large circuits. Figure 3-3 shows the SPICE-compatible circuit, with electrical nodes numbered.

The major adjustments to the model are to provide user control of an applied magnetic field. This is what node 5 and resistor RB are for. When a connected circuit presents a voltage to node 5, that voltage is interpreted as gauss input to the sensor. The resistor to ground is merely to guarantee that the node has a path to ground. This is done for reasons of numerical stability; it doesn't have any function in the circuit other than to make the circuit easier for SPICE to simulate.

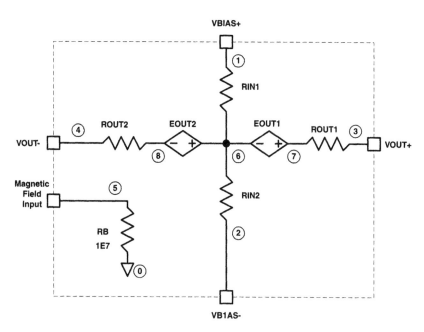

Figure 3-3: Electrical model adapted for SPICE.

This schematic can now be translated into the SPICE language, and packaged as a subcircuit. Because of the number of commercial varieties of SPICE that have evolved over the past few years, we will be using a minimal set of features, so as to provide a least-common-denominator model. In keeping with this philosophy, the controlled sources (EOUT1 and EOUT2) are modeled as multidimensional polynomial functions of V5 and V1–V2. Some versions of SPICE will simply let you specify the source's gain algebraically (e.g., V(5) * (V(1)–V(2))) but this feature is not uniformly supported.

The subcircuit's user I/O ports are:
　　Nodes 1, 2: bias connections (+ and –)
　　Nodes 3,4: output connections (+ and –)
　　Node 5: magnetic field input (1 gauss/Volt)

To finish this example and make this a complete model, we will use some of the parameters of the F.W. Bell BH-200 transducer:

- Sensitivity = 40 μVo/G–V$_{in}$
- R_{in} = 2.5Ω
- R_{out} = 2Ω

The resultant SPICE code is shown in Listing 3-1:

Listing 3-1: Simple SPICE model for BH200 Hall-effect transducer.

```
* EXAMPLE OF SIMPLE HALL-EFFECT TRANSDUCER MODEL FOR SPICE
*
.SUBCKT       HALL1 (1 2 3 4 5)
* HALL EFFECT TRANSDUCER SUBCIRCUIT
* EACH RIN LEG HAS HALF OF 2.5 OHM INPUT RESISTANCE
RIN1    1 6 1.25
RIN2    2 6 1.25
* EACH ROUT LEG HAS HALF OF 2 OHM OUTPUT RESISTANCE
ROUT1   7 3 1.00
ROUT2   8 4 1.00
* EACH SOURCE PROVIDES V5*(V1-V2)*GAIN/2
EOUT1   7 6 POLY(2) 5 0 1 2 0.0 0.0 0.0 0.0 20E-6
EOUT2   6 8 POLY(2) 5 0 1 2 0.0 0.0 0.0 0.0 20E-6
* LOAD FOR MAGNETIC INPUT (AS VOLTAGE) - KEEPS SPICE HAPPY
RB      5 0 1E7
.ENDS HALL1

**** TEST CIRCUIT ****
BIAS WITH 5V
VBIAS 1 0 5
* PROVIDE 1 MEG LOADS FROM OUTPUTS TO GND RL1 2 0 1E6 RL2 3 0 1E6
* DEFINE VMAG AS MAGNETIC FIELD INPUT
VMAG 4 0 0
* CALL HALL TRANSDUCER SUBCIRCUIT
X1 1 0 2 3 4 HALL1
* SWEEP MAGNETIC FIELD FROM -1000 TO +1000 GAUSS IN 20 G STEPS
* AND OUTPUT RESULTS
.DC VMAG -1000 1000 20
.PRINT DC V(4), V(2), V(3)
.PLOT DC V(4), V(2), V(3)
.END
```

This SPICE model provides the following features:

- Input and output resistance
- A control for applied flux, via pin 5
- Output voltage, both as a function of bias voltage and applied field

To make for a simple illustration, we deliberately left a number of relatively useful features out of this model. Temperature coefficients of resistance and sensitivity, for example, are not modeled in the above SPICE input file. SPICE is quite capable of simulating temperature-dependent behavior, provided one goes to the trouble to build an appropriate model. For many, if not most, purposes, however, the level of detail presented in this model will be sufficient for evaluating most of the circuits presented in this chapter. As with any computer model, your actual mileage may vary, depending on how you use (or abuse) it.

3.3 Voltage-Mode Biasing

One of the major dichotomies in the design of a Hall-effect sensor is the mode in which the transducer is biased. It can be driven by either a constant voltage source or a constant current source; both modes have their advantages and their disadvantages. We will first look at circuitry for biasing a transducer with a constant-voltage source.

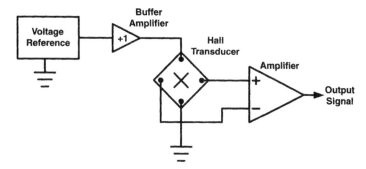

Figure 3-4: Voltage-mode Hall-effect sensor.

Figure 3-4 shows the basic arrangement of a voltage-biased Hall-effect sensor. There is a voltage reference, a buffer, the transducer, and an amplifier. One of the key features of this architecture is that, for most applications, the temperature coefficient of the transducer sensitivity will be sufficiently high that it must be compensated for. This can be done in one of two ways. The first is to make the voltage source temperature dependent so as to obtain a constant output level from the Hall-effect transducer. The second method is to make the gain of the amplifier temperature dependent, so it can compensate for the temperature-varying gain of the transducer signal. Although either method can be made to work, we will examine the case where the drive voltage is held constant over temperature.

A simple but workable bias circuit is shown in Figure 3-5a. A voltage reference, in this case a REF-02 type device, drives an op-amp follower to provide a stable voltage (+5V) to bias the Hall-effect transducer. When constructed from commonly available op-amps, such as the LM324 or TL081, this circuit can provide a few milliamperes of output drive, making it suitable for use with high input resistance transducers ($R_{in} > 1$ kΩ).

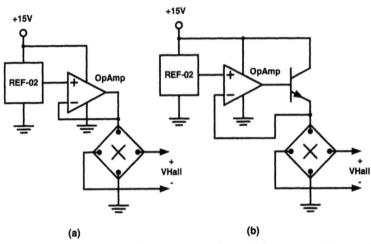

Figure 3-5: Voltage bias circuits for low current (a) and high current (b) transducers.

Bulk and thin-film transducers, however, can often require as much as 100–200 mA for optimal performance. To bias these devices, the driver circuit of Figure 3-5b can be used. The addition of the transistor to the output of the op-amp increases its effective output current capability. When using this circuit, there are a few issues to keep in mind. First is whether or not the transistor has sufficient current gain (referred to as beta or β). The maximum output current able to be drawn from the emitter of the transistor is limited to the maximum output current of the op-amp multiplied by the transistor's beta.

Another potential problem associated with this circuit is that of power dissipation in the transistor. For example, if the transducer draws 100 mA with only 0.2V of bias voltage, it only dissipates 20 mW. The drive transistor, however, if the collector is connected to a 5V supply, will dissipate nearly 500 mW of power. The maximum power dissipated in the transistor must be anticipated in the design, and should be a factor in selecting the transistor, as well as in determining what kind of additional heat-sinking is required, if any.

Finally, in either of the circuits of Figure 3-5, stability can become an issue, especially if the transducer is operated at the end of a length of cable. The addition of parasitic capacitance, or in some cases the additional transistor of Figure 3-5b, can cause the bias circuit to break out into oscillation. Because the stability of a given circuit is dependent on many variables, there is no simple one-size-fits-all fix. The circuit of Figure 3-6 shows one approach that is often useful when driving loads at the end of long cables. This circuit works by providing separate feedback paths from the cable load and the output of the op-amp. The exact values required for the components (R_F, C_F, R_O) surrounding the op-amp will depend on the details of the application and the opamp chosen. For a general-purpose op-amp such as the TL081, setting R_F to 100 kΩ, C_F to 100 pF, and R_O to 100Ω is a good starting point for experimentation in many cases.

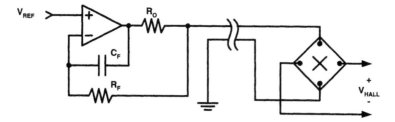

Figure 3-6: Circuit for driving capacitive loads.

All of the above bias circuits result in a small differential Hall output voltage riding on a fairly large common-mode voltage; specifically half the bias voltage. The common-mode signal is the average of the two output voltages, while the differential is the difference. In the case of a Hall-effect transducer, the differential signal is the one carrying the measurement information. While it is possible to measure a small differential signal riding on a large DC common-mode signal, signal recovery is easier if one doesn't have to deal with a large common-mode signal. The circuit of Figure 3-7 solves this problem by symmetrically biasing the transducer. If the transducer's sense terminals are halfway between the bias terminals, providing bias at +V and −V will result in zero common-mode output voltage. The Hall output voltage will in this case swing symmetrically about zero volts, for the case of ideally matched components. In reality this scheme will not completely eliminate common mode voltage from the transducer, but will reduce it to very low levels.

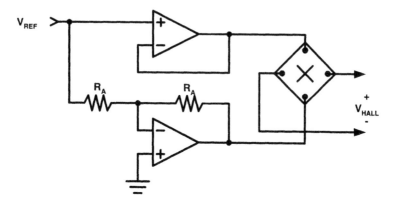

Figure 3-7: Symmetric bias circuit.

One problem frequently encountered when driving a Hall-effect transducer in constant-voltage mode is the voltage drops along long wires. This problem becomes especially acute when the load requires a significant amount of current. While one solution is to simply use thicker wires, this is not always either feasible or desirable. Another option is the use of a force-sense bias circuit. In a force-sense type of bias circuit, four wires are used to provide the bias voltage to the transducer. Two of the wires are used to provide positive supply and return (the force leads), while the remaining two (the sense leads) are used to measure the voltage that is actually being provided to the transducer. Because there is an insignificant amount of current flowing down the sense leads, there is no significant voltage drop along them, and they can be used to accurately measure the transducer bias voltage. Figure 3-8 shows a force-sense bias circuit using a differential amplifier (Gain = 1) to measure the voltage across the transducer. This circuit will impress V_{REF} across the transducer bias terminals even if significant voltage drops (~1V) occur along the force leads.

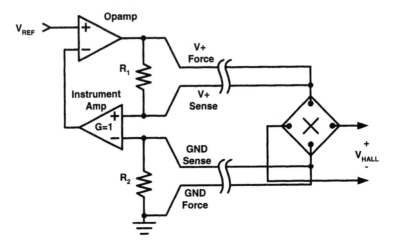

Figure 3-8: Force-sense bias circuit.

Resistors R_1 and R_2 are in this circuit to handle the condition where one of the sense leads is accidentally disconnected. If this situation should occur, the differential amplifier will still measure the voltage applied to the force lead. This prevents the op-amp from overdriving and possibly damaging the transducer in the event of a broken sense connection.

It is also possible to construct a force-sense bias circuit with a single opamp, as shown in Figure 3-9. Good performance in this circuit, however, is highly dependent on the degree to which the values of all the R_A resistors match each other.

Because force-sense techniques are normally employed when the transducer is some distance from the bias supply, stability again becomes an issue. The circuits of both Figures 3-8 and 3-9 would most likely require some modification in order to be successfully used in practice with commonly available op-amps and differential amplifiers.

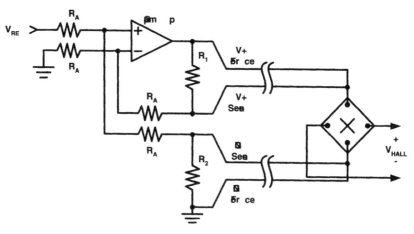

Figure 3-9: Force-sense circuit implemented with single op-amp.

3.4 Current-Mode Biasing

Another way to bias a Hall-effect transducer is to feed it with a constant current; this mode of operation results in a Hall output voltage with a temperature coefficient on the order of 0.05%/°C, as opposed to the 0.3%/°C obtained with constant-voltage biasing. For many applications, the tempco obtained with constant-current biasing is sufficiently low that no additional correction is necessary. An additional advantage is obtained when using long cable runs; because current does not "leak" out of wires to any appreciable degree under most circumstances, current-mode biasing doesn't normally require any kind of force-sense arrangement to be used in the bias supply.

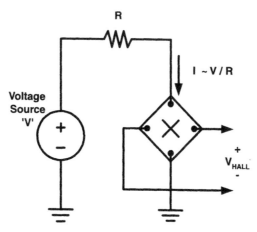

Figure 3-10: Brute-force constant-current bias circuit.

There are a number of ways to construct constant-current bias sources for a Hall-effect transducer. The simplest method is to use a high-value resistor (R) in series with a constant voltage source, as shown in Figure 3-10. While such an arrangement is easy to make and inexpensive, maintaining a stable bias current requires that the voltage source be much greater than the transducer bias voltage. While this arrangement can be useful with bulk-type transducers that often operate with a bias voltage of under a volt, it can require excessively large voltage sources when used with integrated devices that may require several volts of bias.

The use of active electronics allows for the construction of stable current sources that don't require stable high voltage supplies. Figure 3-11 shows three common circuit topologies.

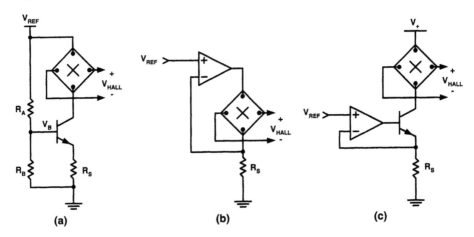

Figure 3-11: Constant current sources. Transistor (a) op-amp (b) op-amp with outboard transistor (c).

The circuit of Figure 3-11a works by setting a voltage at the base of the transistor, approximately determined by $V_B = V_{REF} \cdot R_B/(R_A + R_B)$. This results in an emitter voltage about 0.6V lower, which then sets the current through R_S. The load (transducer) in the collector doesn't significantly affect the amount of current pulled through the collector, assuming that the collector voltage is higher than the emitter voltage so that the transistor is not saturated. The collector current is approximately given by:

$$I_C \approx \frac{1}{R_S}\left(\left(\frac{V_{REF}R_B}{R_A + R_B}\right) - 0.6\right) \qquad \textbf{(Equation 3-1)}$$

This relation will hold substantially true if two conditions are met. The first is that the voltage across R_S is greater than 0.6V; this will minimize the effects of transistor V_{BE} variation (V_{BE} varies both over current and temperature, between individual transistors). The second condition is that $(R_S \cdot \beta) >> ((R_A \cdot R_B)/ (R_A + R_B))$. If this condition is violated the base will load down the R_A/R_B divider network and the output current may be significantly reduced.

The circuit in Figure 3-11b uses feedback to regulate current. By actively measuring the current passing through R_S, and consequently through the transducer, the opamp can adjust its output voltage to obtain the desired current. The current is given simply by V_{REF}/R_S, and if the op-amp has sufficient drive capabilities (current and voltage), the bias current will remain nearly constant over temperature. The output current drive capabilities of this circuit may be increased by adding a transistor to the output of the op-amp, in the manner similar to that shown previously in Figure 3-5b.

A third current source appears in Figure 3-11c. This circuit adds the active feedback control of an op-amp to the circuit of Figure 3-11a. The resultant current is approximately V_{REF}/R_S, with a small error (1–5%) resulting from base current feeding through the emitter and into the sense resistor. Adding feedback makes this circuit much less sensitive to variation in the transistor caused by both device-to-device differences and temperature effects, as compared to the circuit of Figure 3-11a.

One characteristic of all of the above circuits is that the transducer is "floating" with respect to ground, meaning that no terminal is grounded The transducer in Figures 3-11a and 3-11c is essentially "hanging" from the positive supply rail. The transducer of Figure 3-11b is floating at some indeterminate point in between ground and the positive supply rail. A circuit called a *Howland current source* can be used to supply current to a ground-referenced load. Figure 3-12 shows the Howland current source.

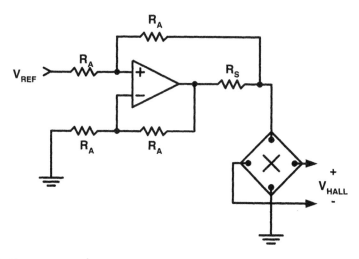

Figure 3-12: Howland current source.

To obtain good performance with a Howland current source, the R_A resistors must be well matched and significantly greater in value than the sense resistor R_S. The output current is given by V_{REF}/R_S.

3.5 Amplifiers

With the transducer properly biased, one obtains a small differential voltage signal from the output terminals, often riding on a large DC common mode signal. The job of the amplifier is to amplify this small differential signal while rejecting the large common-mode signal. The fundamental circuit to perform this task is the differential amplifier (Figure 3-13), also known as an instrumentation amplifier (or in-amp).

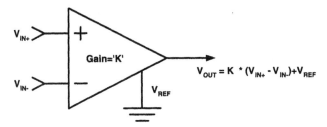

Figure 3-13: Instrumentation amplifier.

A typical differential amplifier has a positive and a negative input terminal and an output terminal. The schematic symbol unfortunately looks a lot like that for an op-amp, sometimes leading to a bit of confusion. Some differential amplifiers have an additional reference input terminal, to which the output voltage is referenced. For most applications, this terminal will be tied to ground. Ideally, the output voltage is the difference of the two input voltages. Because ideal devices are not yet available, you must make some trade-offs among various performance characteristics to get an amplifier that suits your needs. Some of the key parameters for differential amplifiers are:

- Differential Gain
- Gain Stability
- Input Offset Voltage
- Input Bias Current
- Common-Mode Rejection
- Bandwidth
- Noise

Differential gain is the gain by which the amplifier boosts the difference of the input signals. While there are monolithic instrumentation amplifiers that have fixed gains, this parameter is often user adjustable within wide limits, with ranges of 1000:1 commonly available.

Gain stability. One uses an instrumentation amp to get an accurate gain, and this is one of the features that differentiates them from the more common op-amp, which has a very large (>50,000) but not very well-controlled gain. Key gain-stability issues center around initial accuracy (% gain error) and stability over temperature (% drift/°C).

Input Offset Voltage. This is a small error voltage that is added to the differential input signal by the instrumentation amp. It results from manufacturing variations in the internal construction of the amplifier. The offset voltage is multiplied by the gain along with the signal of interest and can be a significant source of measurement error.

Input Bias Current. The inputs of the instrumentation amp will draw a small amount of input current. The amount is highly dependent on the technology used to implement the amplifier. Devices using bipolar transistors in their input stages tend to draw input currents in the range of nanoamperes, while those based on field-effect transistors (FETs) will tend to draw input bias currents in the picoampere or even femtoampere (10^{-15}) range. While FET-input instrumentation amps have lower bias currents than their bipolar counterparts, the input offset voltages are usually higher, meaning that a trade-off decision must be made to determine which technology to use for a given application.

Common-Mode Rejection. While the purpose of a differential amplifier is to amplify just the difference between the input signals, it also passes through some of the common-mode, or average, component of the input signal. The ability of a given amplifier to ignore the average of the two input signals is called the *common mode rejection ratio*, or CMRR. It is defined as the ratio between the differential gain (A_{vd}) and the common-mode gain (A_{vc}) and, like many other things electrical, is often expressed logarithmically in decibels:

$$CMRR = 20\log\left(\frac{A_{vd}}{A_{vc}}\right)$$

(Equation 3-2)

Common-mode rejection ratios of 80–120 dB (10,000–100,000) can be easily obtained by using monolithic instrumentation amplifiers. Additionally, the CMRR for many devices increases as the gain increases.

Bandwidth. Unless you are only interested in very slowly changing signals, you will probably be concerned with the frequency response, or bandwidth, of the amplifier. This is commonly specified in terms of a gain-bandwidth product (GBP). In rough terms, gain-bandwidth product can be defined as the product of the gain and the maximum frequency at which you can achieve that gain. For many types of amplifiers, the GBP is roughly constant over a wide range of frequencies. For example, an amplifier with a 1-MHz GBP can provide 1 MHz of bandwidth at a gain of 1, or conversely only 1000 Hz of bandwidth at a gain of 1000. Figure 3-14 shows how the gain of this hypothetical 1-MHz GBP amplifier varies when set at various gains.

One caveat, however, is that an amplifier doesn't simply block signals past its frequency response; the response gracefully degrades. For precision applications, you will want to choose your bandwidth so that it is at least a factor of 5–10 greater than that of the signal you are interested in. So, for the case of an amplifier with a gain of 1000 amplifying signals with useful information up to about 1000 Hz, you might want to use an instrument amplifier with a GBP of 5 to 10 MHz to preserve signal integrity.

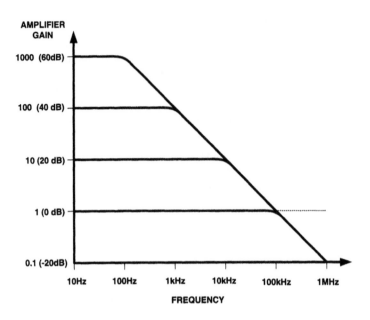

Figure 3-14: Instrumentation amplifier frequency response vs. gain.

Noise. In addition to noise from the transducer, an amplifier will add some noise of its own. Although the sources of amplifier noise are complex and beyond the scope of this text, it can be modeled as a noiseless amplifier, with both voltage and current noise sources at the input, as shown in Figure 3-15. Because the noise from the current source is converted into voltage by the source impedance, it also ultimately appears as voltage noise. For a given input impedance R_S, the total amplifier noise is given by:

$$v_{NT} = \sqrt{\left(v_N\right)^2 + \left(R_s i_N\right)^2}$$

(Equation 3-3)

Noise is specified over a given bandwidth, and is usually given in terms of V√Hz for voltage noise and amperes/√hertz for current noise. As with the case of transducer noise, the larger the bandwidth examined, the more noise that will be seen. (See Figure 3-15.)

Different technologies provide varying trade-offs between the magnitude of the voltage and current noise sources. Bipolar input amplifiers tend to have low voltage noise and high current noise, whereas amplifiers using FET technology tend to have higher voltage noise and lower current noise. The choice of technology is complex and is dictated by both the technical requirements and the economics of an application. As a general rule of thumb, however, bipolar-input amplifiers tend to give better noise performance with low impedance transducers (<1 kΩ) while FET-input devices contribute less noise when used with higher impedance sources. Table 3-1 lists the voltage and noise parameters of a few commonly available op-amps.

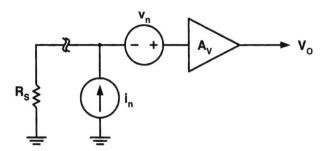

Figure 3-15: Noise model of amplifier.

Table 3-1: Typical noise performance of various operational amplifiers at 1 kHz.

Device	Technology	VN (nV/√Hz)	IN (pA/√Hz)
OP27	Bipolar	3	1
OP42	JFET	13	0.007
TLC272	CMOS	25	Not avail.

3.6 Amplifier Circuits

Although a number of techniques exist for constructing differential amplifiers, one of the simplest is to use an op-amp and a few discrete resistors as shown in Figure 3-16.

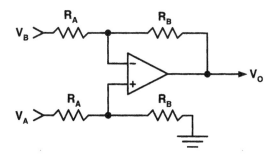

Figure 3-16: Differential amplifier.

This circuit provides an output voltage (with respect to ground) that is proportional to the difference in the input voltages. The output voltage is given by:

$$V_O = \frac{R_B}{R_A}(V_A - V_B)$$ (Equation 3-4)

In the ideal case where the R_A resistors match, the R_B resistors match, and an "ideal" op-amp is used, the common mode gain is zero. In the more realistic case where the resistors do not match exactly, the common mode gain is a function of the mismatch; the greater the mismatch, the higher the common mode gain and the less effective the differential amplifier will be at rejecting common mode input signal. For a differential amplifier constructed with 1% precision resistors, one can expect a CMRR on the order of 40–60 dB (100–1000).

One characteristic of this differential amplifier is that its input impedance is determined by the R_A and R_B resistors. In situations where this circuit is used with a transducer with a comparable output impedance, severe gain errors can result from the amplifier loading down the transducer. One way around this problem is to buffer the inputs with unity-gain followers, as is shown in Figure 3-17a. This circuit can provide very high input impedances, especially when implemented with FET input op-amps.

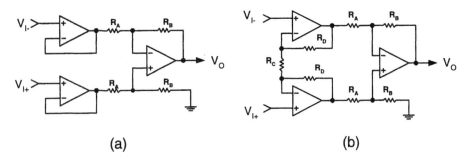

(a)   (b)

Figure 3-17: Three-op-amp differential amplifiers. Simple buffered (a) classical (b).

Figure 3-17b shows the "Classical" three-op-amp instrument amplifier. In this circuit, the input op-amps still provide high-impedance inputs, but they also provide an additional differential gain stage. The common mode gain of this first stage is unity, but the differential gain is given by:

$$A_{VD} = \frac{2R_D}{R_C} + 1$$ (Equation 3-5)

Note that there is only one R_C in the circuit. Making this resistance variable allows for a single-point adjustment of the amplifier's gain.

With the addition of the final stage, the differential gain of the complete instrumentation amplifier is given by:

$$A_{VD} = \frac{-R_B}{R_A}\left(\frac{2R_D}{R_C}+1\right)$$

(Equation 3-6)

The circuit of Figure 3-17b's use of two gain stages offers a number of significant advantages over the circuits of Figures 3-16 and 3-17a. First, it allows for the construction of a higher-gain instrument amplifier for a given range of resistor ratios. Next, by splitting the gain across two stages, it offers higher frequency response for a given gain. Finally, because of the differential gain provided in the first stage, it becomes possible to build amplifiers with much higher levels of common-mode rejection, especially at higher gains.

While there are several other ways to obtain differential amplifiers, perhaps the easiest is to simply buy them; a number of manufacturers presently supply very high-quality devices at reasonable prices. One example of such a device is the AD627. Figure 3-18 shows how the device can be hooked up. Only one external component, a resistor (R_G) to set gain, is necessary for operation. As with any precision analog circuit, local power supply bypassing (in this case provided by capacitor C_B) is usually a good idea.

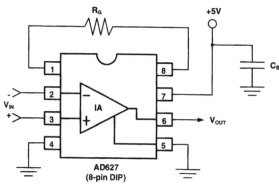

Figure 3-18: External connections for AD627 monolithic instrumentation amplifier.

Aside from component count reduction, the use of monolithic devices offers many performance advantages. Some of the key parameters of the AD627 ("A" version for 5V operation at 25°C) are listed in Table 3-2:

Table 3-2: Key parameters of AD627A instrumentation amplifier.

Parameter	Min	Typical	Max	Units
Input offset voltage	—	50	250	μV
Gain range	5	—	1000	—
Gain error (G=5)	—	0.03	0.10	%

(Continued)

Parameter	Min	Typical	Max	Units
Gain vs. temperature (G=5)	—	10	20	ppm/°C
Input current	—	3	10	nA
Gain-bandwidth product		400		kHz
Common-mode rejection	77	90	—	dB
Input voltage noise (G=100)		37		NV/√Hz

Designing an instrumentation amplifier with this level of performance using dis-crete op-amps and resistors would be a challenging project, especially at a cost less than or equal to that of the AD627. For many applications, using an off-the-shelf monolithic instrumentation amplifier will be the most cost-effective means of achieving a desired level of performance.

3.7 Analog Temperature Compensation

Although it is possible to make voltage sources, current sources, and amplifiers with a high degree of temperature stability, it is difficult to obtain this characteristic in a Hall-effect transducer. While the transducer provides a more constant gain over tem-perature when biased with a constant current (0.05%/°C) than with a constant voltage (0.3%/°C), many applications require higher levels of stability.

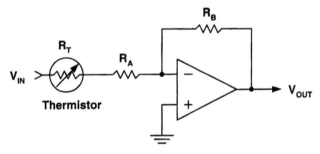

Figure 3-19: Temperature-compensated amplifier.

One method of increasing the temperature stability is to use an amplifier with a temperature dependent gain. Figure 3-19 shows one such implementation based on an op-amp and a thermistor, in an inverting configuration. As the temperature increases, the value of the thermistor decreases, causing the gain to rise. Because the temperature responses of most thermistors are highly nonlinear, it will often be impossible to obtain an exact desired gain-vs.-temperature response for any given combination of resistors and thermistors. The gain for this circuit as a function of temperature is given by:

$$G(T) = \frac{-R_B}{R_A + R_T(T)}$$ (Equation 3-7)

The best that can usually be done is to obtain an exact fit at two predetermined reference temperatures, with some degree of error at temperatures in-between. One procedure for designing this type of circuit, given a set of desired gains and a given thermistor, is as follows:

1) Determine the temperatures at which exact fit is desired (T_1, T_2)
2) Determine the gain at T_1 and T_2 (G_1, G_2)
3) Determine the resistance of the thermistor at T_1, T_2 (R_{T1}, R_{T2})
4) Calculate R_A by:

$$R_A = \frac{(G_2 R_{T2} - G_1 R_{T1})}{G_1 - G_2}$$ (Equation 3-8)

5) Calculate R_B by:

$$R_B = -G_1(R_A + R_{T1})$$ (Equation 3-9)

The above procedure assumes that a positive temperature coefficient of gain is desired and the resistor has a negative temperature coefficient. Note that there will be some cases for which a viable solution doesn't exist (usually indicated by a negative value for R_A or R_B), and others for which the solution may be unacceptable (i.e., resistor values are too low or too high).

Example:
Design a circuit to provide a gain of –9 at 0°C and –11 at 50°C. The thermistor to be used has a resistance of 32.65 kΩ at 0°C and 3.60 kΩ at 50°C. (As a side note, this degree of resistance variation is not at all unusual in a thermistor; the device used in this example is a 10-kΩ (nominal @ 25°C) "J" curve device.)

$G_1 = -9$

$G_2 = -11$

$R_{T1} = 32.65 \text{ k}\Omega$

$R_{T2} = 3.60 \text{ k}\Omega$

                                                                        (Equation 3-10)

$$R_A = \frac{(G_2 R_{T2} - G_1 R_{T1})}{(G_1 - G_2)} = \frac{(-11 \cdot 3.60 \text{ k}\Omega - (-9 * 32.65 \text{ k}\Omega))}{(-9 - (-11))} = 127 \text{ k}\Omega$$

$$R_B = -G_1\left(R_A + R_{T1}\right) = -(-9)(127\ \text{k}\Omega + 32.65\ \text{k}\Omega) = 1436\ \text{k}\Omega \qquad \textbf{(Equation 3-11)}$$

Table 3-3 shows the gain of this circuit for several temperatures, and the departure from a straight-line fit (% error).

Table 3-3: Temperature compensated amplifier performance.

T (°C)	RT (kΩ)	Ideal Gain (straight-line)	Actual Gain (circuit)	Error %
−20	97.08	−8.2	−6.4	−22
−10	55.33	−8.6	−7.9	−8
0	32.65	−9	−9	0
10	19.90	−9.4	−9.8	4
20	12.49	−9.8	−10.3	5
30	8.06	−10.2	−10.6	4
40	5.33	−10.6	−10.8	2
50	3.60	−11	−11	0
60	2.49	−11.4	−11.1	−3
70	1.75	−11.8	−11.2	−5

Although temperature-compensated amplifiers can be useful in certain situations, they suffer from two principal drawbacks. The first is that, as the above example shows, it can be difficult to get the curve one really wants with available components. Although various series and parallel combinations of resistors and thermistors can be used to get more desirable temperature characteristics (with varying degrees of success), this also results in more complex design procedures, with the design process often deteriorating into trial-and-error.

The second drawback is a little more philosophical; the underlying physical mechanism responsible for transducer gain variation will rarely be the same as that underlying the temperature response of your compensated amplifier. In practice this means that you often can't design a compensation scheme that works well over the process variation seen in production. Any given compensation scheme may need to be adjusted on an individual basis for each transducer, or at least on a lot-by-lot basis in production. Since this adjustment process requires collection of data over temperature, it can be an expensive proposition with lots of room for error.

3.8 Offset Adjustment

While some systematic attempts can be made to temperature compensate the gain of a Hall-effect transducer, the ohmic offset voltage is usually random enough so that to make any significant reduction requires compensation of devices on an individual basis. Moreover, since the drift of the ohmic offset will also usually have an unpredictable component, one will often only try to null it out at a single temperature (such as 25°C).

The simplest method of offset adjustment is shown in Figure 3-20; it uses a manual potentiometer to null out the offset of the Hall-effect transducer. The potentiometer is used to set a voltage either positive or negative with respect to the output sense terminal, and a high-value resistor (R_A) sets an offset current into or out of the transducer. It is therefore possible to null out either positive or negative offsets with this scheme.

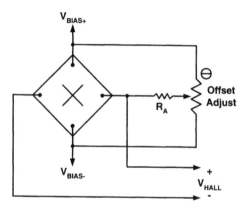

Figure 3-20: Manual offset adjustment using potentiometer.

An adjustment made in this manner is relatively stable over temperature, particularly if the temperature coefficient of resistance of R_A can be made similar to that of the transducer output. Another feature of this trim method is that it can be used regardless of whether the transducer is biased from a constant current or a constant voltage source. The amount of trim current injected through R_A will be proportional to the bias voltage, and will thus track the transducer's ohmic offset over variations in bias voltage.

Offset adjustment can also be made at the amplifier. When performing offset correction in the amplifier circuitry, an important consideration is that the transducer's offset voltage will be proportional to the voltage across its drive terminals. The input offset voltage of the amplifier, however, will be independent of sensor bias conditions. In the case of a transducer being operated from a constant bias voltage, both offsets can be approximately corrected through one adjustment. Because Hall-effect transducers are more typically biased with a constant current source, the transducer bias voltage will change as a function of temperature. There are several approaches to dealing with these two independent offset error sources. The first is to get an amplifier with an acceptably low input offset specification and simply ignore its input offset voltage. A circuit to perform an adjustment of transducer offset is shown in Figure 3-21. This circuit measures the actual bias voltage across the transducer and generates proportional positive and negative voltage references based on the bias voltage. The potentiometer allows one to add an offset correction to the amplified sensor signal. Since this correction will be proportional to the transducer bias voltage, it will track changes in transducer offset resulting from bias variation over temperature.

If a separate offset adjustment were added to correct for input voltage offset of the instrumentation amplifier, the offset adjustment process would need to be performed as a two-stage process. First the amplifier inputs would need to be shorted together and the amplifier offset adjusted, for zero output voltage. Next, the short would be removed, and the Hall-offset adjust would be used to trim out the remaining offset.

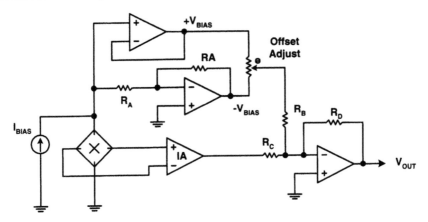

Figure 3-21: Correcting transducer offset at the amplifier.

3.9 Dynamic Offset Cancellation Technique

In addition to removing zero-flux ohmic offset by manually trimming it out, an elegant method exists that exploits a property of the Hall-effect transducer to reduce system offset.

A four-way symmetric Hall-effect transducer can be viewed as a Wheatstone bridge. Ohmic offsets can be represented as a small ΔR, as shown in Figure 3-22a. When bias current is applied to the drive terminals, the output voltage appearing at the sense terminals is $V_H + V_E$, where V_H is the Hall voltage and V_E is the offset error voltage.

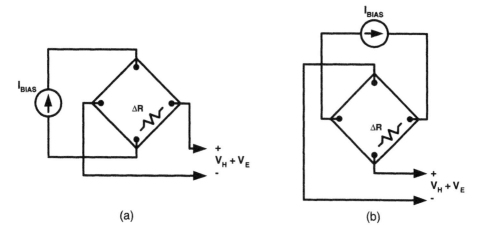

Figure 3-22: Effects of rotating bias and sense terminals on output.

Now consider what happens if we take the transducer and reconnect the bias and sense terminals, as shown in Figure 3-22b. All of the terminal functions have been rotated clockwise by 90°. The sense terminals are now connected to bias voltage, and the former drive terminals are now used as outputs. Because the transducer is symmetric with rotation, we should expect to see, and do see, the same Hall output voltage.

The transducer, however, is not symmetric with respect to the location of ΔR. In effect, this resistor has moved from the lower right leg of the Wheatstone bridge to the upper right leg, resulting in a polarity inversion of the ohmic offset voltage. The total output voltage is now $V_H - V_E$. One way to visualize this effect is to see the Hall voltage as rotating in the same direction as the rotation in the bias current, while the ohmic offset rotates in the opposite direction.

If one were to take these two measurements to obtain $V_H + V_E$ and $V_H - V_E$, one can then simply average them to obtain the true value of V_H. For this technique to work, the only requirement on the Hall-effect transducer is that it be symmetric with respect to rotation.

It is possible to build a circuit that is able to perform this "plate-switching" function automatically. By using CMOS switches, one can construct a circuit that can rotate the bias and measurement connections automatically. Such a circuit is shown in Figure 3-23. An oscillator provides a timing signal that controls the switching. When the clock output is LOW, the switches A,B,E, and F close, and switches C,D,G, and H open. This places the transducer in a configuration that outputs $V_H + V_E$. When the clock goes HIGH, switches C,D,G,H close, while the remaining switches open. This outputs $V_H - V_E$.

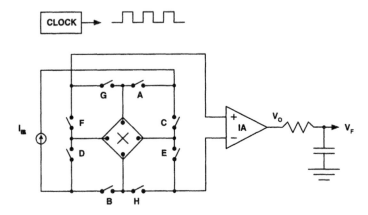

Figure 3-23: Switching network and output filter for auto-nulling offset voltage.

The output signal produced by the transducer and this periodic switching network next needs to be amplified and averaged. While there are several ways of averaging the signal over time, the simplest is through the use of a low-pass filter. Examples of signals present at various stages of this network are shown in Figure 3-24.

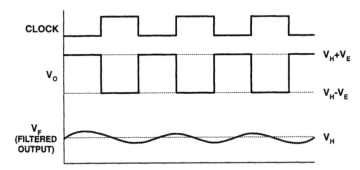

Figure 3-24: Auto-nulling circuit waveforms.

This technique is used, with suitable modifications, in many modern Hall-effect integrated circuits. It is often referred to by the terms *plate-switching*, *auto-nulling*, and *chopper-stabilization*. When properly implemented, it can reduce the effective ohmic offset of a given transducer by nearly two orders of magnitude. Additionally, since the offset is being cancelled out dynamically, it is very temperature stable. This technique can also be employed in the construction of the digital-output Hall-effect ICs, resulting in highly sensitive devices with low and temperature-stable operate (turn-on) and release (turn-off) points.

Chapter 4

Integrated Sensors:
Linear and Digital Devices

While it is certainly possible to build one's own Hall-effect sensors using transducer elements and discrete signal-processing components, using techniques such as those described in Chapter 3, it is usually unnecessary to do so for most applications. Because Hall-effect sensors are widely used, a number of semiconductor companies include a variety of integrated Hall-effect sensors among their standard product offerings. These devices contain a silicon Hall-effect transducer as well as the bias, amplification, and signal-processing circuitry needed to obtain an easy-to-use output signal. These circuits come in integrated circuit packages that can be soldered into printed circuit boards, or to which discrete wires can be attached. Some examples of the packages in which Hall-effect ICs are manufactured are shown in Figure 4-1.

Figure 4-1: Hall-effect ICs in various packages. *(Courtesy of Melexis USA, Inc.)*

Adding electronics to the transducer allows integrated Hall-effect ICs to provide several types of functions for the user, in addition to merely providing a linear output signal. These integrated sensors can be roughly divided into four categories:

- Linear output devices
- Digital-output threshold-triggered devices (switches and latches)
- Speed sensors
- Application-specific devices ("none-of-the-above")

Linear output sensors provide a continuous output that is proportional to magnetic field strength.

Switches and latches provide a digital output that actuates and resets when an applied magnetic field exceeds and drops below preset threshold levels. These types of integrated Hall-effect sensors are by far the most widely used.

Speed sensors (also called *geartooth sensors*) consist of one or more Hall-effect transducers combined with application-specific circuitry designed to detect the passage of moving ferrous targets such as gear teeth.

Application-specific devices are arbitrarily defined by the author as those that do not fall into one of the above three categories. These devices are developed to meet the needs of a specific application, typically those of a specific customer. These devices tend to enter a manufacturer's "standard" product line when exclusivity agreements with the original customer expire.

Of all the integrated Hall-effect sensors manufactured today, the most commonly applied devices in terms of unit volume are switches, latches, and linear-output devices. These simple devices are the fundamental building blocks implementing Hall-effect based applications. This chapter will describe the characteristics of these particular integrated sensors.

4.1 Linear Sensors

From a user-level standpoint, a linear Hall-effect sensor provides an output voltage proportional to applied magnetic field. Integrated linear Hall-effect sensors employ many of the techniques described in Chapter 3, suitably adapted for use on a monolithic IC. By integrating the support circuitry required to interface to a Hall transducer, significant advantages are obtained in terms of size, power consumption, and cost.

Linear Hall ICs are typically based on one of three overall architectures: linear, transducer switching, and digital.

Figure 4-2 shows the typical organization of a *linear architecture* linear Hall IC. The key components are a bias source, the Hall sensor, a differential amplifier, and a regulator circuit for consistently biasing the amplifier. The amplifier is often designed so that its gain and offset can be permanently adjusted at the time of manufacture, typically through the use of either fusible links or through "zener zapping." For good stability, this type of sensor is almost always made in a precision bipolar process, resulting in fairly large die sizes. The Allegro Microsystems A1301 is an example of this type of device.

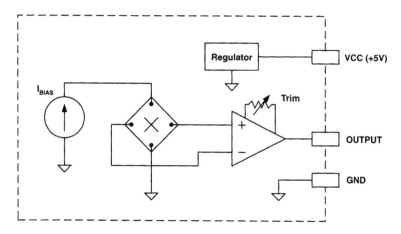

Figure 4-2: Linear Hall-effect IC.

Transducer switching architectures, often referred to as *chopper-stabilized, auto-zeroing*, or *auto-nulling*, utilize the dynamic offset compensation techniques described in Chapter 3. A typical auto-nulling Hall IC is shown in Figure 4-3.

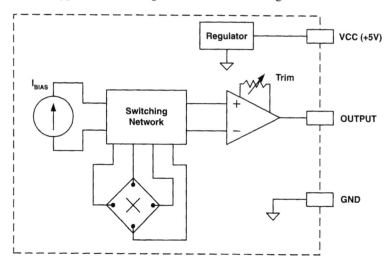

Figure 4-3: Auto-nulling linear Hall-effect IC.

Because auto-nulling techniques can greatly reduce the effects of amplifier input offset in addition to the offsets resulting from the Hall-effect transducer, CMOS amplifier circuits, with their notoriously poor performance, can be effectively used for signal processing. One advantage of implementing this type of sensor in CMOS is that the switching network is straightforward to implement with CMOS transistors. Although it is possible to use bipolar devices to perform the necessary switching functions, such circuits can be complex and difficult to implement. Additionally, since circuitry can be built more densely in CMOS processes than in precision bipolar processes, a CMOS Hall-effect sensor can incorporate more features than a bipolar one of comparable cost. For this reason, auto-nulled linear Hall-effect sensors such as the Melexis MLX90215 can incorporate user-programmable gain, offset, and temperature compensation, all for a modest additional cost over that of nonadjustable bipolar linear devices.

The use of CMOS technology can be taken to even more of an extreme in the case of a linear Hall-effect sensor that uses digital signal processing techniques. In this device, the signal from the Hall transducer is amplified and fed into an analog–to-digital converter (ADC). Gain and offset correction are then performed digitally, through binary multiplication and addition (Figure 4-4). The result is then converted back into an analog signal with a digital-to-analog converter (DAC).

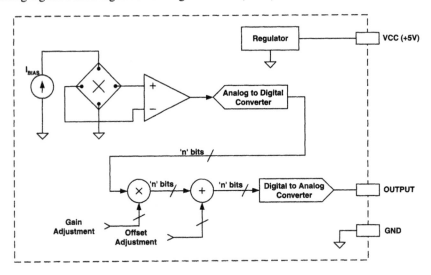

Figure 4-4: Digital linear Hall sensor

This digital signal processing (DSP) architecture offers many implementation advantages. Because the sensor's signal processing chain now consists almost entirely of digital logic, it can be fabricated with very high-density processes, reducing manufacturing cost. Another benefit of this approach is that gain and offset can now be stored in tables that are accessed as a function of temperature, allowing for very precise calibration of the device. One example of this type of device is the Micronas HAL805.

While there are quite a few ways in which to implement a linear Hall sensor, a device's performance can be usefully described by a few common characteristics. Some of the more important of these characteristics are described in the following sections.

4.2 Linear Transfer Curve

The most basic characterization of a linear transducer is its transfer curve. This function defines the relationship between the magnetic field sensed by the device and its corresponding electrical output. An example of a transfer curve from a "typical" Hall sensor IC is shown in Figure 4-5. Some of the defining features of this curve are:

1) **Zero-flux Intercept.** This defines output voltage under conditions of zero magnetic field. This point is variously referred to as *quiescent voltage out* (Q_{VO}), and *zero-field* offset.

2) **Slope** dV/dB. This feature defines the sensor's gain, or sensitivity. Because the slope is not constant over the transfer curve and is difficult to directly measure at a given point, manufacturers typically pick two points (B_1, B_2) within the device's "linear" range, often symmetric about zero flux, and calculate average sensitivity over this range as $S = (V_2 - V_1)/(B_2 - B_1)$.

3) **Output Saturation Voltages.** Unless unusual design measures are taken, the signal output range of a device operating from a single 5V power supply will be limited between 0V and 5V. The minimum and maximum limits between which the output signal can vary are known as the saturation points. Since most linear sensors are designed to operate from a single 5V supply, negative saturation (V_{SAT-}) is usually specified with respect to ground, while positive saturation (V_{SAT+}) is specified as how close the output can swing to the positive supply rail. Because many popular analog-to-digital converters have input ranges of 0–5 volts, most manufacturers offer "rail-to-rail" outputs on many of their newer linear Hall-effect sensors, for ease of interface. While the output voltages on these devices can't really swing all the way to the supply rails, they can often come within a few tens of millivolts if they are not required to drive an excessive load.

4) **Useful linear range.** Even though a device's output can have a wide range of operation before it saturates, the linearity may significantly degrade long before saturation is reached. The useful linear range is both a function of the device and the nonlinearity error a given application can tolerate. For most modern Hall sensor ICs, the linearity error is well below 1% over the majority of their specified sensing range. The limits of the effective linear range are denoted as B_{SAT+} and B_{SAT-} on the transfer curve. Note that, in the example shown, the output voltage continues to increase and decrease past these limits, but becomes significantly more nonlinear. In addition to being a function of a given sensor's design, the useful linear output range can also be a function of the electrical load the sensor must drive.

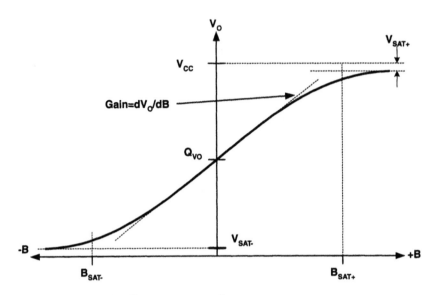

Figure 4-5: Linear Hall-effect sensor transfer curve.

4.3 Drift

Unfortunately, a transfer curve is only valid under a given set of environmental conditions. For many Hall-effect sensor applications, the most germane environmental influence is usually temperature. While it is certainly possible to characterize a device with a family of transfer curves, each corresponding to performance at a given temperature, the more typical approach is to provide characterization of only the sensitivity and zero-flux output voltage over temperature.

Simplifying the temperature characterization in this manner makes it easier to specify screening parameters for testing individual ICs. Devices can be readily screened against both absolute limits and drift limits for both zero-flux output voltage and sensitivity.

One important point to keep in mind is that low-temperature drift for either zero-flux offset or sensitivity (gain) does not necessarily imply low temperature drift for the other. This is particularly evident in the case of auto-nulled devices, which can exhibit offset drifts on the order of a few gauss over their rated temperature range. Some of these same devices can also exhibit sensitivity drifts on the order of ±5% over that same temperature range, a drift comparable to that of many older bipolar Hall-effect sensors. Conversely, an old-style bipolar Hall sensor can have excellent sensitivity stability, but poor offset drift characteristics. The moral of this story is that you need to determine which device characteristics are important for your application and then to make sure that the device selected provides adequate performance in those areas.

4.4 Ratiometry

One of the primary design objectives when developing a Hall-effect sensor is to develop a device that is minimally susceptible to environmental influences. There is, however, one environmental factor, other than magnetic field, to which many devices are deliberately made very sensitive. Many linear Hall-effect sensor ICs are designed so that their sensitivity and zero-flux offset are linear functions of the supply voltage. A device with this property is referred to as having a ratiometric output.

A ratiometric Hall sensor has a zero-flux output voltage set at some fraction (often ½) of the supply voltage, and a sensitivity that is proportional to the supply voltage. This means that if you vary the supply voltage by 10%, the zero-flux output will increase by 10%, and so will the sensitivity. While increasing the device's sensitivity to power supply variation may seem to be counterproductive, a ratiometric sensor can actually be very useful as a system component.

Consider the case in which the output of a Hall-effect sensor is fed into an analog-to-digital converter (ADC) shown in Figure 4-6. In this example, a voltage regulator provides both the power supply for the Hall-effect sensor (V_{CC}) and a reference voltage (V_{REF}) for the ADC.

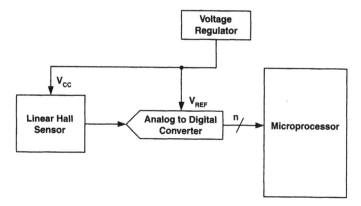

Figure 4-6: Example of system benefiting from a ratiometric Hall-effect sensor.

In this example, first let us consider the case of a nonratiometric sensor, in which the Hall-effect sensor provides a fixed sensitivity regardless of supply voltage. If the voltage regulator's output voltage were to rise, the Hall-effect sensor would still provide the same sensitivity as it would at a lower regulator voltage. Because the full-scale span of the ADC would increase with rising regulator voltage, the sensor's output would "fill" less of the ADC's range, resulting in smaller reported measurements by the ADC.

One solution might be to insist on using a highly stable voltage regulator to prevent this type of error from occurring. To implement a Hall-effect sensor with a sensitivity that is supply-voltage independent, however, requires that some kind of precision voltage or current reference be put inside the sensor IC. In this situation, there are now really two separate references that must be stable—one inside the sensor and one outside. It is always more difficult to maintain precision across two references than one.

Now let us consider the case in which the gain and offset of the Hall sensor is proportional or ratiometric to its power supply. If the voltage regulator's output should increase, the sensor's offset voltage and sensitivity will increase proportionately. The ADC's input span will also increase by the same amount. The sensor's output range will map to the ADC's input range in the same way regardless of small variations in regulator voltage. This means that, for a given magnetic field, the ADC will output the same binary code independent of the reference voltage.

Ratiometric output Hall sensors, however, are not without their drawbacks. To make absolute measurements, a stable voltage source is needed at some point, if only for initial calibration. Another problem is that of power supply noise rejection. For the case of a device exhibiting "ideal" ratiometric behavior, you would expect to see approximately half of the power supply noise to appear at the output. The power-supply noise susceptibility of actual devices will vary depending on the details of their implementation.

4.5 Output Characteristics

The output of a linear Hall-effect sensor can be usefully modeled by a voltage source (v_o) in series with a resistor (r_o), as shown in Figure 4-7.

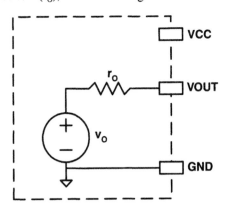

Figure 4-7: Equivalent model of linear Hall-effect sensor output.

Most contemporary devices employ negative feedback in their output amplifier stages, mainly because this improves output linearity. Negative feedback also provides an additional benefit of low dynamic output impedance, often a few ohms or less. The low output impedance of such devices simplifies the design of interface circuits. When these types of sensors are used to drive load impedances greater than a few thousand ohms, gain errors resulting from loading effects tend to be minimal.

A low output impedance, however, does not imply that a given sensor can sink or source a large amount of output current. Typical output currents tend to be limited to a few milliamps. If a load circuit sinks or sources more than the rated current from a device, it can seriously degrade the device's effective gain, zero-flux offset, and linearity. In some cases, excessive output current can also damage the device.

4.6 Bandwidth

The bandwidth, or frequency response, of a linear system can be described by a Bode plot, which is a graph of gain and phase lag vs. frequency. For both traditional and practical reasons, the frequency and gain axis are usually expressed in either logarithmic units (dB for gain) or on a log scale (for frequency). A Bode plot corresponding to what is called a "first-order, low-pass system" is shown in Figure 4-8.

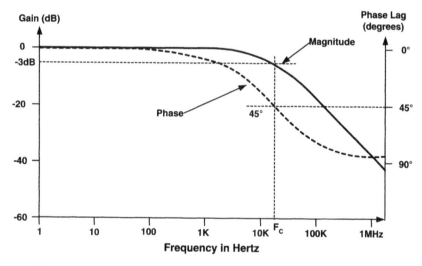

Figure 4-8: Bode plot of first-order linear system.

Some of the key features of the system described in the Bode plot are:
1) **Corner frequency** F_C, often called the –3 dB point. At this frequency the value of system gain or sensitivity is only $1/\sqrt{2}$ (0.707) of its value at DC (zero frequency).
2) **Attenuation rate**. Beyond the corner frequency the sensitivity of a first-order system rolls off at the rate of –20 dB per decade of frequency. Another way to look at it is that the response drops a factor of 10 for every 10× increase in frequency.

3) ***Phase shift***. At the corner frequency, the system will delay a sinusoidal input signal by 45°. As one increases the signal frequency, this phase delay increases asymptotically to 90°.

While measurements such as the corner frequency describe a system's response to sinusoidal stimuli, they also provide some insight into how a device behaves in response to other stimuli. One common example is the system's response to an input in the form of an abrupt step, usually called the *step response*. The time that a first-order system needs to settle to within 37% of its final value (called the system *time constant*, represented by τ) is given by $1/(2\pi f_c)$. Figure 4-9 shows the step response of a first-order, low-pass system.

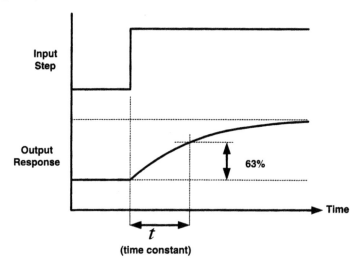

Figure 4-9: Step response of first-order, low-pass system.

The amount of time this system needs to settle to within a specified error bound can be expressed as a function of τ, as shown in Table 4-1.

Table 4-1: Error vs. settling time, ideal first-order system.

Settling time in τ units	% Error
1	36.8
2	13.5
3	5.0
4	1.8
5	0.6

A typical linear Hall-effect sensor has a corner frequency in the range of 10 kHz–25 kHz. This corresponds to a τ of about 6 to 16 microseconds.

Although a linear Hall-effect sensor is much more complex than a first-order system, especially if it employs auto-nulling techniques, using a first-order model is a useful approximation for many applications. Manufacturers of linear Hall-effect devices will usually publish "typical" corner frequency values in the data sheets for their devices.

4.7 Noise

Noise can be loosely defined as any signal you are not interested in seeing. All linear electronic systems generate some amount of internal noise, which they add to their output signal. This is also true for linear Hall-effect sensors. Manufacturers typically specify the internal noise of their linear Hall-effect sensors in one of two ways. The first is to express it as a peak-to-peak measurement, when looked at over a specified frequency range. The other method is to express the noise as an RMS (root-mean-square) equivalent noise voltage, also over a specified frequency range. Comparing peak-to-peak noise measurements to RMS noise measurements is difficult, especially if they are specified over different frequency ranges. Fortunately, detailed noise measurements for a given device can be readily performed with common electronics lab equipment such as a spectrum analyzer or an RMS voltmeter. A spectrum analyzer is an especially useful tool because it tells you how much noise appears at any given frequency. Knowing the frequency distribution of the noise often allows it to be at least partially filtered out.

4.8 Power Supply Requirements for Linear Sensors

Modern linear Hall-effect sensors have modest power supply requirements. In contrast to many analog circuits, linear Hall-effect sensors almost always operate from a single positive power supply (often +5V or +12V) for operation. Current consumption through the power supply lead is often less than 10 mA. A power supply decoupling capacitor, with a value typically ranging between 0.001 μF and 0.1 μF is often connected across the sensor's power supply terminals to reduce noise and guard against spurious operation. Power supply decoupling is especially important for ratiometric sensors, as power supply noise can readily couple through to the output.

While many integrated Hall sensors (switches, latches, speed sensors) have significant protections against conditions of supply voltage overload and reversal (reverse battery protection), many linear devices, especially those providing rail-to-rail output drive capabilities, don't. Linear devices tend to be much more susceptible to electrical damage through the power supply terminals than their digital counterparts. Special care should be taken to ensure that devices are not exposed to power supply conditions that exceed their absolute maximum ratings.

4.9 Temperature Range

Linear Hall-effect sensors are available over a number of operating temperature ranges. The most common are:

Commercial:	0° to 70°C
Industrial:	–40° to 85°C
Automotive:	–40° to 125°C

Devices are also sometimes available over other ranges. Those intended for under-the-hood (engine compartment) automotive applications are often rated to operate at temperatures up to 150°C.

4.10 Field-Programmable Linear Sensors

Sensor-based products are often required to meet some kind of parametric performance requirements. For example, in the case of a position sensor, the device may be required to provide a predictable change in voltage output corresponding to a specified change in positional input. The individual components of a Hall-effect-based sensing system all have some finite unit-to-unit variation associated with them. The magnetic components used in the assembly are also subject to variation in material properties and dimensions, resulting in unit-to-unit variation in magnetic output. The Hall-effect sensors will also show some unit-to-unit variation in their sensitivities and offsets. The "stack-up" of all these sources of variation can make it exceedingly difficult to implement a design that meets a tight set of performance objectives.

While it is often possible to perform a unit-by-unit adjustment of the magnetic and mechanical portions of a sensor to meet a performance target, these kinds of operations may be difficult to realize in a cost-effective manner in a production environment. Another option is to add external circuitry to the sensor to allow for gain and offset adjustment. Again, however, this can add considerable cost to a product.

One recently available solution that has emerged to address this problem is the field-programmable linear Hall-effect sensor. Many types of Hall sensor ICs are adjusted electronically on a one-time basis at the time of manufacture. The idea of the field-programmable linear Hall-effect sensor, however, is to allow the end-user to perform this adjustment, after the device has been assembled into the final product. A block diagram of a typical field-programmable linear sensor is shown in Figure 4-10.

A typical field-programmable sensor provides several adjustable features:

Coarse and Fine Gain – A coarse gain setting is often provided to select a general range of sensitivity (mV/gauss) for the sensor, while the fine range setting is used for trimming the sensitivity to exactly what the user wants. For example, one may select a coarse range of 1–2.5 mV/gauss, and then use the fine range to adjust the gain to 1.50 mV/gauss. A combination of coarse and fine gain settings is typically provided because

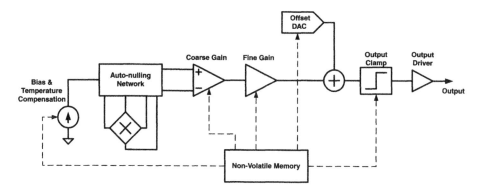

Figure 4-10: Field-programmable linear hall sensor.

it is usually simpler and less costly to implement two separate gain stages, one with wide dynamic range and the other with high resolution, than it is to implement a single stage with comparable dynamic range and resolution.

Output Offset Voltage – A DAC is often provided to adjust the output offset voltage. While offset voltages in modern auto-nulled linear Hall sensors are typically fairly low to begin with, this feature is often more useful to deliberately bias the output. For example, if the device will only be sensing fields of one polarity, a negative offset can be set so the device will output zero volts for zero field. Any applied field would then drive the output up towards the positive supply rail (e.g., +5V). For a typical nonprogrammable device, in comparison, the output for zero flux is usually a voltage halfway between ground and the positive supply rail (e.g., ~2.5V). In this application, the ability to offset the output effectively increases the device's output signal swing. Another application for adjusting the output voltage offset is to offset the effects of any baseline applied field. As an example, consider a position sensor assembly where the magnetic field varies from 100 gauss to 200 gauss from the beginning to the end of its travel range. By offsetting the Hall-effect sensor's output voltage to compensate for the 100-gauss starting point, the sensor's output voltage could be 0 volts at the beginning point of travel. By appropriately adjusting the gain, one could also make the sensor report 5V at the endpoint of travel.

Output Clamp – For some applications it is desirable to limit the sensor's output voltage swing to a narrow range. This is common when designing sensors into systems that must be able to identify failures and attempt to "fail safe." As an example, consider a sensor operating from a +5V supply and providing a valid output ranging from 0V to 5V. In the event the sensor became disconnected from a system relying on its output, that system would read zero volts, and have no way of detecting the fault. Similarly, if the sensor output line was to be shorted to the +5V sensor supply line, the monitoring system would see +5V and also have no way of determining that there was a fault condition. A clamping function can be used to limit the sensor's output voltage excursion to a known voltage range, such as 1V to 3V. Figure 4-11 shows the effects of a clamping function on the sensor's output voltage.

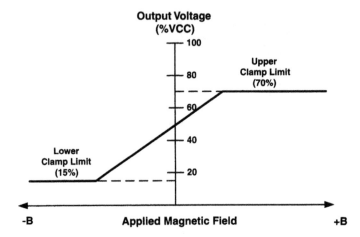

Figure 4-11: Effect of clamping function.

Because the sensor limits its valid output signals to levels located between the clamp limits, any signal falling outside these limits can then be interpreted as a fault condition by downstream electronics, and handled appropriately. Field-programmable sensors often provide the user with the ability to either select from several predetermined clamping limits, or provide the ability to set the clamp limits at arbitrarily selected voltage levels.

Adjustable Temperature Compensation – By making the transducer bias current a function of temperature, it is possible to make the sensor's overall gain also vary with temperature. While temperature coefficients are usually things to be minimized in the case of traditional nonprogrammable linear sensors, adjustable temperature coefficients can be a valuable feature in a user-programmable device. The primary application for setting the sensor's temperature coefficient for gain is in matching its response so as to compensate for the temperature coefficients of magnetic materials it may be used to sense. For example, NdFeB magnets typically have a negative temperature coefficient on the order of –0.1%/°C, meaning that their magnetic field output goes down as temperature goes up. By setting the sensor's temperature coefficient to offset that of the magnet (e.g., +0.1%/°C) it is possible to reduce the system's overall temperature coefficients to very low levels.

4.11 Typical Linear Devices

Table 4-2 lists a few typical linear Hall-effect linear ICs from various manufacturers, as well as their nominal room-temperature gains.

Table 4-2: Selected linear hall ICs.

Device	Manufacturer	Nominal Gain (mV/G)	Comments
A1301	Allegro	2.5	Continuous time, Not auto-nulling
A1373	Allegro	Note 1	Auto-nulling
MLX90251	Melexis	Note 1	Auto-nulling
HAL805	Micronas	Note 1	DSP

Note 1: Parameter is user-programmable.

4.12 Switches and Latches

Many magnetic sensing applications merely require knowing if a magnetic field exceeds a given threshold; a detailed measurement of the field is unnecessary. Because Hall-effect sensors are frequently used in this type of application, manufacturers have found it worthwhile to incorporate threshold-sensing electronics into their integrated sensors. The resulting sensor ICs provide a digital On/Off output. A block diagram of a "digital" or threshold-sensitive Hall-effect sensor is shown in Figure 4-12.

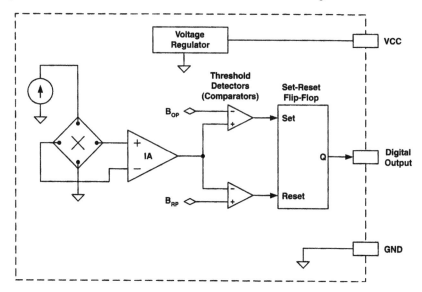

Figure 4-12: Functional block diagram of threshold-sensitive Hall IC.

A digital Hall-effect sensor essentially takes a linear device (regulator, bias circuit, Hall transducer, and amplifier) and adds a threshold detector and a digital output driver.

As in the case of linear devices, auto-nulling circuitry is often employed to improve stability. When the value of the sensed magnetic field exceeds an arbitrary turn-on threshold (often referred to as B_{OP}, for B, Operate Point), the upper comparator activates the SET input on the flip-flop, forcing it into the ON state. The flip-flop subsequently drives the output ON. When the value of the magnetic field drops below an arbitrary turn-off threshold (often referred to as B_{RP}, for B, Release Point) the lower comparator activates the RESET input of the flip-flop, forcing it into the OFF state. The output is then driven into the OFF state. When the value of the magnetic field is between the two limits, the flip-flop maintains the state to which it was last set or reset. This effect is called hysteresis, and prevents the output of the device from oscillating between the ON and OFF states when the magnetic field is near a threshold. The amount of hysteresis (B_H) for a given device is determined by $|B_{OP} - B_{RP}|$. Figure 4-13 shows two ways of viewing the behavior of a digital-output Hall sensor. If one ramps the magnetic field (B) up and down again as a function of time (Figure 4-13a), the device turns on when the field exceeds B_{OP}, and the device turns off when the field is reduced below B_{RP}. Digital Hall-effect devices are characterized by ramping a known magnetic field up and down and noting the flux densities (B) at which the device turns on and turns off.

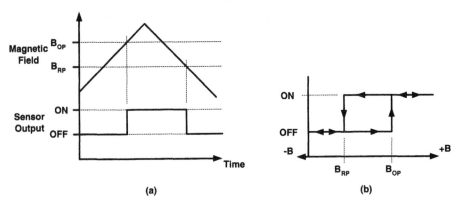

Figure 4-13: Digital sensor reaction to ramped field (a), and transfer function (b).

Another way to visualize the device's behavior is to plot the transfer curve of output vs. magnetic field, as is shown in Figure 4-13b. Because the transfer function varies depending on the direction in which one sweeps the magnetic field, one ends up with two overlapping curves, with the width of the "eye" between the up and down curves indicating the amount of hysteresis.

One important point to remember is that the digital sensors described in this section react to magnetic field as an algebraic quantity. This means that a "positive" field is always interpreted as greater than a "negative" field. While there are devices that respond to the absolute value of the applied field, they constitute a minority of available devices and will be discussed in a later section. For various historical reasons, most contemporary digital Hall-effect sensors are designed so that a South magnetic pole presented to their front surface is interpreted as a positive field, while a North magnetic pole similarly presented is interpreted as a negative field.

4.13 Definition of Switch vs. Latch

The points at which B_{OP} and B_{RP} are set have a profound effect on how the sensor behaves in responses to magnetic stimuli. Assuming that $B_{OP} > B_{RP}$, the three cases are:

1) $B_{OP} > 0, B_{RP} > 0$
2) $B_{OP} > 0, B_{RP} < 0$
3) $B_{OP} < 0, B_{RP} < 0$

In case #1, where both B_{OP} and B_{RP} are positive, the device remains normally off when no magnetic field is applied, turning on only when a sufficiently strong, positive field is sensed. When the field is removed the device returns to the OFF state. This device is called a Hall-effect switch.

In case #2, where B_{OP} is positive and B_{RP} is negative, the device can be turned on by a sufficiently strong positive magnetic field, but only can be turned off with a sufficiently strong negative field. When the magnetic field is removed, the device remains in whatever state it presently is in. Because of this memory, or latching effect, this device is called a Hall-effect latch. Case #3 follows on the next page.

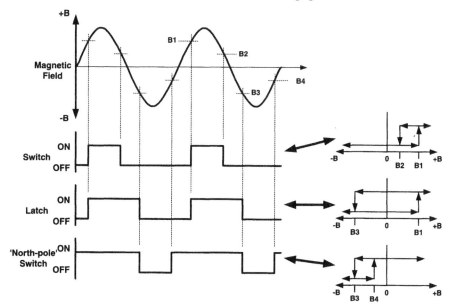

Figure 4-14: Behavior of switch (a), latch (b), north-pole switch (c).

Finally, in case #3, where both B_{OP} and B_{RP} are negative, the device remains normally on in the absence of applied field, and can only be turned off by a sufficiently strong negative field. Devices that exhibit this behavior are commonly referred to as North-pole switches, at least in the case when positive magnetic field is defined as South pole.

Figure 4-14 illustrates the behavioral differences among the three cases described previously. Now that we have described how digital Hall-effect sensors behave, and the concepts of operate and release points (B_{OP}, B_{RP}), and hysteresis (B_H), we will describe some of the other characteristics used to specify the devices.

4.14 Switchpoint Stability

The parameters B_{OP}, B_{RP}, and B_H all will drift over variations in ambient temperature and power supply voltage. For modern Hall-effect sensors, switchpoint (B_{OP}, B_{RP}) variation as a function of power supply voltage is usually minimal. Temperature-induced variations in the switchpoints is more of a concern in most applications. Hall-effect IC manufacturers will usually list limits within which these parameters must remain under a given range of conditions on their device data sheets.

Less commonly will a manufacturer place limits on the drift of these parameters. This is mainly because to do so requires the ability to track individual devices through multiple temperature tests, and use the characterization data from each of these tests to perform a final sorting operation. This type of testing, where one tracks unit-identity of individual devices through multiple tests, is much more expensive to perform than a more traditional test sequence, in which one rejects out-of-spec parts after each individual test operation.

Because the qualitative behavior of a digital part depends on the relationship of B_{OP} and B_{RP}, close attention must be paid to their performance over temperature. To have a latch turn into a switch at a temperature extreme, or vice-versa, can have catastrophic consequences on the system in which the sensors are employed.

4.15 Bipolar Switches

A bipolar switch, is not, as one might first believe from the name, a device that turns ON in response to either pole of a magnet. A bipolar switch is a Hall-effect IC for which the ranges of both the B_{OP} and B_{RP} specifications include zero. This means that, while the device is guaranteed to turn ON when the sensed field exceeds a specified positive threshold and is guaranteed to turn off when the sensed field drops below a specified negative threshold, it may or may not latch when the field is removed. It could be functionally either a latch, or a switch, or even a north-pole switch, the exact function varying on a unit-to-unit basis, even for devices of the same type, and manufactured together in the same lot. Because bipolar switches can exhibit qualitatively different behavior on a unit-to-unit basis, one should carefully review designs in which they are used to ensure that the possible operating-mode variation will not cause spurious operation or malfunctions in the system in which they are incorporated.

4.16 Power Supply Requirements for Digital Sensors

Digital devices tend to be more forgiving of power supply variation than their linear counterparts. Contemporary devices operate over wide supply ranges, with 4 to 24 volts not uncommon. Devices that were designed for automotive applications can often survive significant electrical overvoltage conditions, both negative and positive, on the power supply lead. Devices with a reverse-battery protection feature are designed to withstand continuous and significant (typically exceeding –12V) negative voltages on the power supply lead. While relatively common for digital Hall-effect devices, features such as reverse-battery protection are almost unheard of for most other ICs; applying negative supply voltage to most logic or analog ICs will usually destroy them in short order. Devices designed for automotive applications may also incorporate features to protect them from the short, high-energy transients that commonly occur in vehicle electrical systems.

Most digital Hall sensors draw a relatively small amount of supply current (usually <10 mA) in normal operation. Because the supply current can vary a few milliamperes depending on the output state, particularly in bipolar devices, placing a small bypass capacitor (0.01 – 0.1 μF) across the device's power supply terminals is highly recommended. Because of the on-off nature of the device's behavior, excessive noise on the power supply line can manifest itself as an unstable output state, with output oscillations occurring at times when the device is transitioning between the on and off states, as shown in Figure 4-15a. Connecting a capacitor between the device's output line and ground is NOT the way to fix this problem. While you may be able to damp output oscillations with a sufficiently large capacitor on the output, this tactic is only a superficial fix and does not address the root problem. In some cases a capacitor on the output may actually damage the device due to large current spikes flowing into the output when the device discharges the capacitor. The most effective way to reduce supply-noise oscillations is to appropriately filter and bypass the power supply lines. The simplest method of supply bypassing is to put a capacitor between the sensor's positive power terminal (VCC) and ground, as shown in Figure 4-15b.

In the case of latch-type sensors, good power-supply bypassing is absolutely critical for proper operation. Because a latch is expected to "remember" its present state in the absence of field, it is especially susceptible to having its state altered by spikes, noise, or "dropout" (brief voltage reductions) on the power supply line. One symptom of an improperly bypassed latch is to see its output occasionally return to a preferred state (ON or OFF) without any apparent magnetic stimulus to cause it to switch. In the author's experience, many types of latches seem to be more susceptible to being spuriously flipped by electrical noise on the power-supply lines when they are operated at supply voltages near their lower operating supply voltage limits.

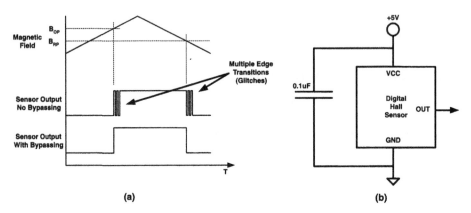

Figure 4-15: Effects of excessive supply-line noise on digital Hall-effect sensor output (a) and use of bypass capacitor (b).

4.17   Output Drivers

Overwhelmingly, the most common output driver found on digital Hall-effect sensors is the NPN open-collector output, shown in Figure 4-16a. When the device turns ON, it biases the output transistor, sinking current into the output. For devices fabricated with CMOS technologies, an N-channel MOSFET replaces the NPN bipolar transistor, and the output is referred to as *open-drain* (Figure 4-16b).

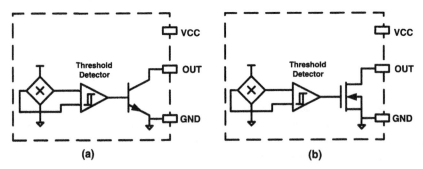

Figure 4-16: Open-collector (a) and open-drain (b) outputs.

Open-collector and open-drain outputs are popular because they are easy to inter-face to. The addition of a single pull-up resistor allows one to interface with both TTL and CMOS logic, as well as with most common microcontrollers. While a few devices incorporate the pull-up on-chip, most require it to be added externally. Forgetting to add the external pull-up resistor is the cause of a great many applications problems. This is probably the single most common problem encountered when first starting to use digital Hall-effect sensors. If a pull-up resistor is not added, you will not get a voltage-output signal out of the device. When viewed with a voltmeter or an oscilloscope, you will see either no signal out, or a small (<0.5V), highly corrupted signal. When operating with a properly sized pull-up resistor, the voltage output signal from an unloaded digital Hall-effect sensor will appear as a clean square wave with a low value near ground and a high value near the voltage to which the pull-up resistor is tied.

While an open-collector output behaves like a switch to ground, it is a solid-state device, and has a few characteristics that must be taken into account when designing interface circuits.

1) **Maximum on-state sink current** (I_{OMAX}). This is the maximum amount of current the output can sink in the on state without incurring damage.

2) **Maximum off-state output voltage** (V_{OMAX}). In the off state, this is the maximum voltage the output can tolerate without breaking down (turning into a short-circuit).

3) **Output off-state leakage current** (I_{OLK}). In the OFF state the output will sink a small amount of leakage current, typically on the order of nanoamperes or microamperes. This parameter defines the maximum amount of this leakage current under a given set of conditions.

4) **Saturation voltage** (V_{sat}). When the output is switched on, and is conducting a specified amount of current, the voltage at the output will not be quite zero. Saturation voltage defines the maximum value for the ON state output voltage for a given current.

5) **Rise and Fall time** (T_r, T_f). These parameters define how fast the output will transition between the ON and OFF states. These characteristics should NOT be confused with the amount of time the device needs to respond to a magnetic field. The response time of a digital device is a complex function of the packaging and circuit architecture used to implement the device. While rise and fall times are often measured in tens or hundred of nanoseconds, the actual response time is often measured in microseconds.

4.18 Typical Digital Devices

Table 4-3 lists a few devices representative of the vast (and I do mean vast!) assortment of digital Hall-effect ICs available today.

Table 4-3: Magnetic characteristics of various digital Hall-effect sensors.

Manufacturer	Device	Type[1]	"Typical" values[2] (gauss)		
			BOP	BRP	BH
Allegro Microsystems	A3121	Sw	350	245	105
	A3134	BSw	8.5	−19	27
	A3141	Sw	100	45	55
	A3240	Sw	35	25	10
	A3280	L	22	−23	45
Melexis	US3881	L	50	−50	90
	US5881	Sw	280	230	43
Micronas	HAL505	L	140	−140	280
	HAL506	Sw	72	50	27

Note 1: Sw=Switch, Bsw=Bipolar switch, L=Latch
Note 2: Typical values from manufacturer's data sheets for test conditions as specified by manufacturer.

Chapter 5

Interfacing to Integrated Hall-Effect Devices

Integrated Hall-effect sensors almost always incorporate enough signal-processing and support circuitry to provide an immediately useful output signal, be it a proportional voltage or a digital switched output. There are situations, however, when the electrical interfaces provided by the sensor do not meet the requirements of a particular application. In these cases, the designer must provide some additional circuitry to bridge the gap between the sensor's output and the inputs of the system to which it is being interfaced.

5.1 Interface Issues—Linear Output Sensors

Because the output of a linear Hall-effect sensor is a voltage proportional to magnetic flux density, it offers the highest degree of flexibility in interfacing to an outside system. By providing the ability to measure magnetic field, a linear sensor is a building block with which one can begin to implement nearly any kind of magnetic sensing function required. Potentially an enormous number of types of interfaces can be implemented; four of the more common types are:

1) Offset and sensitivity adjustment
2) Line-driver circuits
3) Output thresholding
4) Analog-to-digital converter (ADC) interface

The following sections describe circuits to perform these interface functions.

5.2 Offset and Gain Adjustment

A common interfacing situation occurs when the output offset and span of a sensor do not match the input offset and span of some other electronic circuit that needs that sensor's input. While some of the newer linear Hall-effect sensors offer user-adjustable gain and offset controls on-chip, circumstances will still arise in which what is available will not satisfy a given set of requirements. One example would be where a signal swing of ±10V is required. No common integrated linear Hall-effect sensor provides an output that swings below 0V, or has an output range that spans 20V (−10V to +10V). If such an output is needed, it requires the addition of some custom circuitry.

In systems where one has the luxury of operating from ± split power supplies, such as +5/−5V or +15/−15V, the circuit of Figure 5-1a provides trim for both sensitivity and offset. The sensitivity gain is given by $(R_4 \times R_6)/(R_5/R_1)$. The range of offset adjustment as referred to the output of op-amp A_2 is $\pm (V_{ref} \times R_4 \times R_6)/(R_2 \times R_5)$ volts (within the limitations of the op-amps' range of output voltage swing). With some types of op-amps, it may be necessary to keep $R_4 \geq R_1$ and $R_6 \geq R_4$ to keep the circuit from oscillating. In many cases, where a large amount of gain is not needed, you can set $R_5 = R_6$, providing the output stage with a gain of −1, and simplifying the total gain to R_4/R_1 and the offset adjust range to $V_{ref} \times R_4/R_2$. When implementing this circuit with "typical" op-amps, suitable values for the resistors will range from roughly 1000Ω to 1 MΩ.

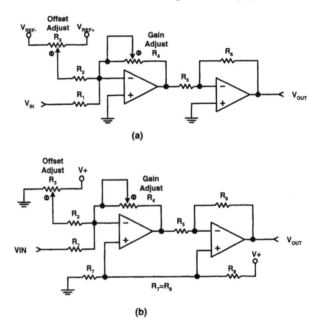

(a)

(b)

Figure 5-1: Offset and gain adjustment circuits for split voltage supply systems (a) and single voltage supply systems (b).

Note that this circuit has two separate amplification stages and requires two op-amps. This is because the first-stage amplifier configuration is inverting; inputting a positive signal results in a negative output signal. Adding another inverting gain stage flips the polarity back to positive. In this circuit, the offset adjustment is applied "upstream" in the signal-processing chain before the gain adjustment. The proper order in which to adjust this circuit is to first adjust the gain to the desired level, through resistor R_4 and, once a satisfactory gain has been achieved, then to adjust the output offset voltage, using R_3.

In many cases, one doesn't have the luxury of operating from a split voltage supply, and has to make the system work from whatever is available, often from a single +5V or +12V supply. The circuit of Figure 5-1a can be modified for single-supply operation by providing the op-amps with a "virtual ground" somewhere between the positive supply and ground. For a 5V system, 2.5V is often a good choice for the virtual ground value. The simplest way to make a virtual ground is by connecting a resistive divider between the positive supply and ground. In this circuit, this function is performed by resistors R_7 and R_8. Placing a small bypass capacitor at this virtual ground point is often helpful in improving the circuit's stability and reducing the output noise.

In cases where cost or space is an issue, it is also possible to perform gain and off-set adjustments with fewer components. An example of a single-supply gain and offset adjustment circuit that uses only a single op-amp is shown in Figure 5-2. In this circuit, gain and offset adjustment are very much interdependent. Adjusting the offset will also change the gain. This is because the gain of the circuit is partially dependent on the impedance looking into the wiper of potentiometer R_3, which ranges from zero when it is at either of the ends (minimum or maximum adjustment) to $R_3/4$ when it is in the middle (where R_3 is the end-to-end resistance of the potentiometer). Needless to say, having the gain and offset interdependent can make this circuit truly a joy to properly adjust. If you choose a value for R_3 that is much smaller than that of R_2, the gain of the circuit can be approximated by $R_2/(R_1 + R_2)$, and the output offset adjustment range by $5V \times (R_1/R_2)$ (for 5V operation). Note that when you adjust the wiper on R_3 toward the positive supply, the output voltage will go down.

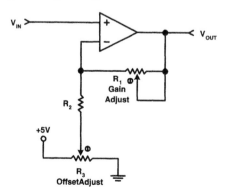

Figure 5-2: Offset and gain adjustment with a single op-amp.

Depending on the requirements of your application, it is also possible to "hardwire" the gain and offset adjustments on the previous circuits by replacing the potentiometers and variable resistors with fixed resistors. If fixed offset and gain adjustments are all that are needed, this is often a desirable thing to do. Variable resistors are among the least reliable electronic components available; fixed-precision resistors offer orders of magnitude higher reliability, for orders of magnitude lower cost. Additionally, manually adjusting a variable component is an expensive and error-prone operation to perform in even a medium-volume production environment. If you don't absolutely need the adjustability, don't use variable resistors in your circuits.

5.3 Output Thresholding

Although there is an enormous variety of switch and latch-type Hall-effect sensors available, providing a multitude of available B_{OP} and B_{RP} points, there may be occasions where you need a device with specific B_{OP} and B_{RP} values. One option is to use a programmable switch-device such as the Allegro Microsystems A3250. In cases where your requirements fall outside the range of available fixed and programmable devices, it is possible to build your own switches and latches around a linear output Hall sensor. This approach allows you nearly complete freedom in setting B_{OP}, B_{RP} and hysteresis. The primary drawbacks are that your discrete version of a switched device will be considerably more expensive than a comparable integrated device and will take up much more space. If your application allows for these increased cost and space requirements, however, then the do-it-yourself approach can be viable.

One method of building a digital sensor out of a linear one is through the circuit of Figure 5-3. This is similar to the conceptual model of a switch presented in Chapter 3, except with a few more implementation details included. When the output of the linear Hall-effect sensor exceeds the B_{OP} threshold, it causes a HIGH (5V) condition at the output of comparator U_{2A}. This HIGH condition causes the flip-flop (U_3) to latch into the HIGH state. When the output of the linear device drops below the B_{RP} threshold, it causes a HIGH at the output of comparator U_{2B}, latching the flip-flop into a LOW state. The hysteresis is the difference between the B_{OP} and B_{RP} points.

The circuit in Figure 5-3 allows for B_{OP} and B_{RP} to be set completely independently, with B_{OP} normally set higher than B_{RP}. For many applications, however, one will be interested in being able to set a single threshold, with only a very small amount of hysteresis. In the presence of a low level of hysteresis, the B_{OP} and B_{RP} thresholds may be nearly indistinguishable. A simple circuit that provides a single threshold with a small amount of hysteresis is shown in Figure 5-4.

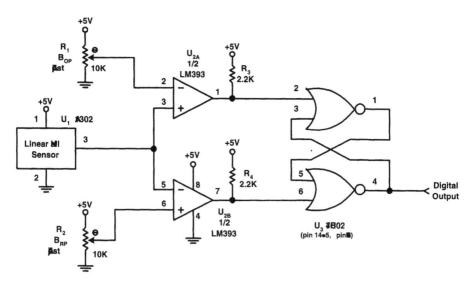

Figure 5-3: Discrete implementation of switch.

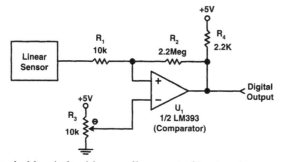

Figure 5-4: Threshold switch with a small amount of hysteresis.

The comparator in this circuit works in the same way as those of the previous circuit, outputting a HIGH condition when the voltage at its positive input is greater than the voltage at its negative input. The network comprising R_1 and R_2 provides positive feedback in this circuit. This means that when the output switches HIGH, it forces the comparator's positive input to also move to a slightly higher voltage. Conversely, when the output switches low, it pulls the positive input down with it. The effect is analogous to the snap in a light switch on an electric lamp—when you push it past the halfway point in either direction, a spring in the switch tends to pull it the rest of the way. The amount of this shift, which is the threshold detector's hysteresis, is approximately given by $V_S * R_1/(R_1 + R_2)$, where $V_S = 5V$ in this circuit. For the values of resistors and V_S

shown, the hysteresis will be about 24 mV. For a linear sensor such as the Allegro A3515, with a sensitivity of 5 mV/gauss, 24 mV of electrical hysteresis would translate into just under 5 gauss of magnetic hysteresis.

Despite its apparent simplicity, this circuit has several drawbacks. Because the amount of hysteresis is dependent on the swing of the comparator's output, anything you do to load down that point will reduce the effective hysteresis. Another issue is that the resistance of R_2 must be significantly greater than that of R_4; otherwise, the hysteresis feedback will load down the output all by itself. If one takes precautions to ensure that the loading on the circuit's output is controlled, this circuit can provide reasonably well-behaved and predictable performance.

5.4 Interfacing to Switches and Latches

One might think that, because the output of a switch or a latch is digital, the whole job of interfacing is already complete. While it is a straightforward matter to interface the output of a typical Hall-effect switch to a digital input, there are also many other, not so obvious ways that digital Hall-effect sensors can be used.

5.5 The Pull-Up Resistor

As mentioned earlier, most digital Hall-effect switches and latches provide an open-collector or open-drain type of output. The output therefore behaves like a switch to ground, and does not output any voltage on its own. In order to derive a useful voltage-level signal, one must add a pull-up resistor, as shown in Figure 5-5. One simple but common question that arises from the necessity of the output pull-up resistor is how to select an appropriate resistor.

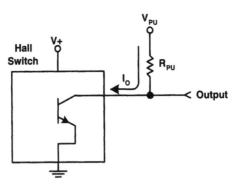

Figure 5-5: Output pull-up resistor.

Only a few pieces of information are required to select and size a pull-up resistor. The first is V_{PU}, or the voltage to which the pull-up resistor is to be tied, which is not necessarily the same as the supply voltage for the sensor. The second piece of information is the current (I_{ON}) that should be flowing through the Hall sensor's output when it is in the ON state. This current is not necessarily the maximum current the device can handle (I_{Omax}), but is determined by the application (limited, of course, to less than I_{Omax}!). Just because a Hall-effect sensor can handle 20 mA through its output doesn't mean that the output current must be 20 mA. Sometimes, indeed most of the time, sizing a pull-up resistor for a lower current level makes sense. In any case, for a given V_{PU} and a desired output current I_O, the resistance value of a suitable pull-up resistor is given by:

$$R_{PU} \approx \frac{V_{PU}}{I_O}$$ **(Equation 5-1)**

For the case of $V_{PU} = 12$ V, and $I_O = 10$ mA, an appropriate pull-up resistor value would be 1.2KΩ.

The second piece of information is the power rating of the resistor. If you do not pick a resistor with a power rating greater than its actual power dissipation, the circuit may experience reliability problems in the field. Again, this is a simple calculation, with power dissipated given by:

$$P = \frac{\left(V_{PU}\right)^2}{R_{PU}}$$ **(Equation 5-2)**

To continue the example above, a 1200Ω pull-up resistor attached to a 12V pull-up supply would dissipate ($12^2/1200$) or 0.12 watts. While this would not pose a problem for a through-hole ¼W resistor, it would be an overload condition for an 0805-type surface-mount device, which is typically rated for 1/10W of power dissipation.

The above analysis for power dissipation assumes that the Hall sensor is constantly in the ON state, and that the resistor is constantly dissipating power. In many applications, the Hall sensor will be turning on and off, and the average power dissipation may be considerably lower than the case in which the sensor is constantly on. While one could size the resistor based on a presumed "duty cycle," it is much better practice to design the circuit to be able to deal with the worst-case power dissipation—i.e., that the sensor is always on and dissipating power in the resistor. Additionally, it is also generally good design practice not to operate electronic components at their maximum power ratings. For example, one would probably not want to select a 1/8W (0.125W) resistor to serve in a place where it could dissipate 0.12W on a continuous basis; a larger resistor, say 1/4W, might be a better choice. In addition to considering a device's power dissipation and ratings, one must also consider the effects of factors such as ambient operating temperature, airflow, and secondary packaging. Selecting component power ratings can be a complex process, well beyond the scope of this text.

Because electronic components tend to be particularly rugged devices, however, a device being driven at a significant thermal overload may work for a while. This means that, instead of failing immediately in the factory, where it merely becomes scrap, an assembly with overloaded devices may make its way out to a customer, where it can become a field-return or a field-service call, or worse. With the introduction of small surface-mount components with limited power dissipation (1/16W for a 0603-sized resistor), component power ratings have become an issue that must be taken very seriously, even in circuitry not normally perceived as being "high-power."

5.6 Interfacing to Standard Logic Devices

Interfacing digital Hall-effect output devices to most modern digital logic devices is extremely straightforward. Here are some tips for interfacing to some of the more common logic families in present use.

Bipolar TTL – (74xx, 74Sxx, 74LSxx, 74Fxx) – connect the pull-up resistor to the logic power supply (Vcc, +5V). Because very little current (< 1 mA) is required to pull a TTL input to logic HIGH (>2.4V), you can use a large pull-up (2. 2kΩ – 4.7 kΩ) for lower system power consumption. While a TTL logic input driven from an open-collector output may sometimes seem to work fine with NO pull-up resistor, this is a bad situation from a reliability standpoint, and should be avoided.

CMOS – (74HCxx, 74HCTxx, CD4xxx) – The pull-up should be connected to the positive voltage supply (VDD) for the logic. For 74H and 74HC families, VDD can range from about 3V–6V, and from 3V–15V for CD4xxx family devices. Because CMOS digital inputs look like a capacitor, very high pull-up resistors (100 kΩ) can be used in situations where one does not especially care about the rise time of the signal. If the resistor is made too large, however, the rise time on the sensor output signal may become unacceptably long. In the case of CMOS logic, if you forget to put in the pull-up resistor, the circuit will work fine as long as the Hall-effect switch's output is ON (pulling the output low to ground) but when it turns off, the input to the CMOS logic gate will float and assume a highly unpredictable value. This type of design error can be very difficult to troubleshoot. One symptom of this problem is that the output stays low when a DVM or an oscilloscope is hooked to the offending logic input, but becomes spurious when the test instrument is removed.

Microprocessor inputs – While most microprocessor input ports look like CMOS inputs (and usually are CMOS inputs nowadays), several variations are occasionally encountered. Built-in pull-up resistors are fairly common on many microprocessor input ports, and can save you from having to add external pull-up resistors. One problem, however, that comes up when trying to interface digital Hall-effect sensors to a microprocessor or microcontroller is that many of the I/O lines are bi-directional, meaning that they can be either inputs or outputs, depending on how they are configured by the software running on the microprocessor. It is important to ensure that there are no tug-of-war situations in which the output of the Hall sensor and an output of the microcontroller both try to control the logic level of a single line. When such a contentious condition exists, one device will lose, your circuit will not work reliably, and in extreme cases hardware damage may result.

5.7 Discrete Logic

It is not uncommon for a sensor assembly to incorporate some amount of logical decision-making capability. In the future this will become increasingly common. In most cases, to implement simple logic functions one would use small-scale integrated logic circuits, such as 74HCxx family devices or programmable logic devices (PLDs). For more sophisticated decision-making capabilities, a small microcontroller might be used.

To perform a few very simple logical operations, such as inverting the polarity of an output, it sometimes makes more sense in a design to use logic functions made from discrete components, such as resistors, transistors, and diodes. There are two reasons for taking this approach. The first is cost. If a logic function can be implemented with a few discrete components, it can actually be more cost effective than one implemented with an integrated circuit. The second reason is that most integrated logic families have a limited operating voltage range. For a sensor assembly that must operate over a 4–24V supply voltage range, using off-the-shelf logic circuits usually requires a voltage regulator to provide a suitable stable power supply (often 5V or 3.3V) for the logic circuits. For both of these reasons there are situations where it often makes sense to design one's own logic functions from discrete components.

The design of robust discrete-transistor logic is a nontrivial task. Many factors, such as operating voltage range, power consumption, immunity to electrical noise, and switching speed must be considered in the course of designing the circuits. When you use a standard logic IC, these issues have already been addressed by the logic manufacturer. When you design your own logic from discrete components, you are on your own. The example circuits that follow should be considered a starting point and not final designs; as simple as they are, they may require significant modification to meet the exact requirements of a given application.

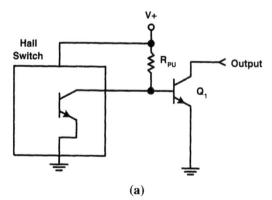

(a)

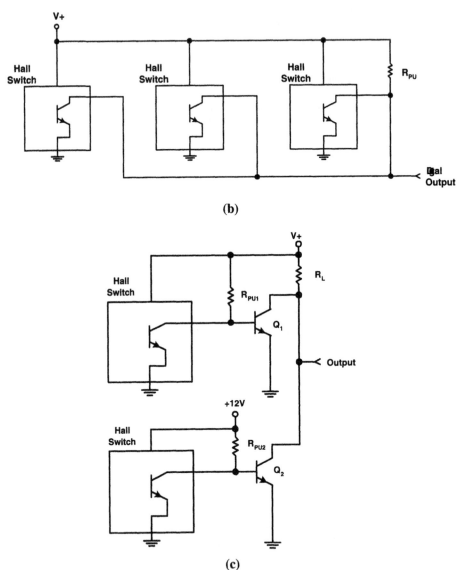

(b)

(c)

Figure 5-6: Discrete logic circuits. Logic NOT (a), Logic OR (b), Logic AND (c) polarity inverter.

Logic NOT Output

This circuit (Figure 5-6a) converts a "normally open" (OFF) open collector output into a "normally closed" (ON) open collector output. When the Hall-effect sensor (U_1) is

OFF, current flows through the pull-up resistor (R_{PU}) into the base terminal of transistor Q_1, turning it ON. Conversely, when the Hall-effect sensor is ON, it diverts current from Q_1, leaving it OFF. This function is called *logic NOT* or *logic inversion*.

Wired OR Output

Open collector outputs can be tied together using a single pull-up resistor, as shown in Figure 5-6b. The resulting circuit is called a *wired-OR* configuration (also sometimes called a *wired AND* in logic design). When one or more of the Hall-effect sensors turns ON, the output is pulled low (ON). This function is known as *logic OR*.

AND function

Sometimes you need to know when *all* of a group of sensors are on. This function is called *logic AND*. The circuit of Figure 5-6c shows one way to do this. It works by combining logic NOT functions with the wired OR function. If a given sensor (U_1) is ON, it causes the associated transistor at its output (Q_1) to be OFF. If both Q_1 and Q_2 were OFF, as would result from both sensors (U_1 and U_2) being ON, then the output will be pulled high to +12V through R_L. If one needs an output that is switched to ground when all of the sensors are on, one can use the output signal shown to drive the base of an additional transistor (not shown) to obtain this switching function.

This circuit also suggests the upper limit of complexity for which it is probably worth designing logic from discrete transistors. This circuit contains three resistors and three transistors. When the costs of the components and the costs of stuffing them onto a circuit board are considered, using an integrated AND gate is likely to be a more economical option, despite possible requirements for additional support circuitry.

5.8 Driving Loads

Hall-effect sensors are sometimes used to control various types of lamps and electromechanical devices that require a significant amount of power to operate. This section will describe some simple circuits for interfacing with these types of devices.

5.9 LED Interfaces

Most digital Hall-effect sensors are capable of sinking up to 20 or 25 mA of output current safely. This also happens to be a more-than-adequate amount of current for driving many small light-emitting diodes (LEDs). This makes it easy in most cases to directly drive an LED from a digital Hall sensor. Figure 5-7a shows a circuit in which the LED lights up when the Hall-effect sensor is ON. Figure 5-7b, on the other hand, shows a circuit in which the LED illuminates when the Hall-effect sensor is OFF; when the sensor is ON in this case, it shunts current away from the LED. In either case, the major design parameters are the desired operating current for the LED and the voltage at which one needs to operate. Because a red LED typically has a forward voltage drop of about 1.8V (the voltage drop depends on the variety of LED—for example, blue LEDs usually have voltage drops of around 3.5 to 4.0V) you need to take this into account to get the right operating current, especially when operating at lower (5V) supply voltages. An appropriate resistor value (when using our typical red LED) can be estimated by

$$R_1 \approx \frac{V_S - 1.8}{I_{LED}} \qquad \textbf{(Equation 5-3)}$$

where V_S is the supply voltage and I_{LED} is the desired operating current. Another issue is selecting the appropriate power dissipation for the resistor. In the case of the first circuit (Figure 5-7a), the resistor only dissipates power when the LED is ON, and the approximate dissipation is given by $(V_S - 1.8V)^2/R_1$, (again, where 1.8V is the LED's voltage drop). In the case of the circuit of Figure 5-7b, the resistor dissipates power all the time, with a maximum dissipation of V_S^2/R when the LED is OFF. One odd "feature" of this circuit is that it actually draws more current when the LED is not illuminated.

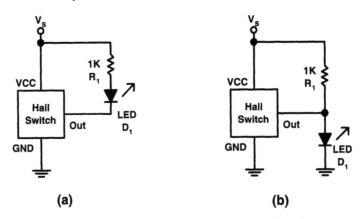

Figure 5-7: Normally OFF LED driver (a), normally ON LED Driver (b).

5.10   Incandescent Lamps

A low-voltage (3–12V) incandescent lamp can provide much more light output than an LED, but also requires much more current to produce that light. From a circuit standpoint, an incandescent lamp is a resistor. Unlike a typical resistor, however, an incandescent lamp will draw a brief surge of up to several times its normal operating current when turning on because the filament resistance is lower when the lamp is cold. This surge is called the cold-filament current or inrush current, and can be more than

10 times the lamp's normal operating current. When designing a lamp driver, one must use transistors that can handle these short high-current inrush spikes without self-destructing. Two simple circuits for controlling small lamps (12V, < 1A current) from the outputs of digital Hall-effect sensors are shown in Figure 5-8.

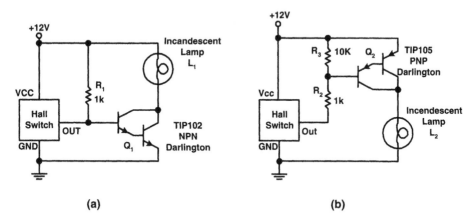

(a)   (b)

Figure 5-8: Incandescent lamp drivers using bipolar Darlington transistors. Inverting (a), and noninverting (b).

These lamp drivers use what is known as Darlington transistors because they provide a much greater current gain (typically >1000) than that of a typical bipolar transistor (typical gains from 50 to 200). This ensures that the ≈10 mA of available base current will completely switch the transistor ON while carrying a load current of up to several amperes. Figure 5-8a shows an inverting driver (Lamp ON when sensor OFF), while Figure 5-8b shows a noninverting driver (Lamp ON when sensor ON). R_3 is used in this circuit to ensure that the transistor turns completely OFF when the output of the sensor is OFF (and not sinking current).

The circuits shown above do show a few nonideal behaviors. The price paid by using a Darlington transistor is high output saturation voltage. The Darlington's collector-emitter saturation voltage, even when switched ON completely, will never drop below about 1V. Aside from the reduction in voltage to operate the lamp (resulting in less light output), this also results in significant power dissipation in the transistor. For example, a 1Ampere current in either of the circuits of Figure 5-8 will dissipate approximately 1W in Q_1, possibly requiring the transistor to be heat-sinked.

Another, and typically better, approach is to use a power MOSFET as the lamp driver. Figure 5-9 shows an inverting lamp driver using a power N-channel MOSFET (lamp is ON when sensor is OFF). When switched completely ON, the particular MOSFET shown in Figure 5-9 has an ON resistance of about 0.2Ω, and will only dissipate 200 mW with a 1A load current. MOSFETs with lower ON resistances are also available, down to values in the range of a few milliohms. Since no current flows into the gate terminal of a MOSFET, it is possible to use MOSFET drivers to switch large load currents, while being controlled directly from the output of a digital Hall-effect sensor.

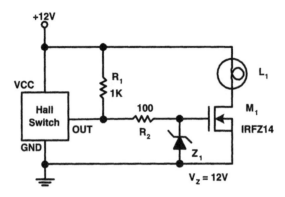

Figure 5-9: Intverting lamp driver using power MOSFET.

MOSFETs, however, have their own set of idiosyncrasies. The first is that since they are voltage controlled devices, you need a significant voltage swing at their gates terminal to completely turn them ON. Depending on the device and the operating conditions, this usually ranges anywhere from 3 - 12 volts. If R_1 were tied to a 5 volt supply, this circuit would either work marginally (the lamp would be dim and the MOSFET would get hot in the ON state), or not at all. One remedy to this problem is to select a MOSFET that is designed for operation at low-gate drive voltages.

Another characteristic of MOSFETS is that their gates are sensitive to damage from even momentary over-voltage conditions. For this reason, it is common to put a zener diode or other device to clamp the maximum gate-source voltage excursions. In this example a 12V zener diode was selected to clamp the gate-to-source voltage excursion. Some modern power MOSFETs, however, have maximum gate-to-source voltage ratings of as little as ±6V, and must be protected accordingly.

A final characteristic of MOSFETs to be discussed here is gate capacitance. The gate of a power MOSFET looks like a large capacitor, typically in the range of 100–10000 pF. It is often advisable to put a small resistor (R2) in series with the gate to both limit the turn-on and turn-off current spikes, and to help prevent oscillation. Despite a few additional and nonobvious design issues to consider, for many power switching applications MOSFET transistors are a better solution than bipolar transistors (Darlington or otherwise).

5.11 Relays, Solenoids, and Inductive Loads

While resistive loads, such as incandescent lamps are relatively benign (once one accounts for the issue of cold-filament inrush current), inductive loads such as relays and solenoids are another story. If you were to drive even a small relay (e.g., 12V, 50-mA coil drive) with one of the lamp drivers described above, the circuit would probably work for a few ON-OFF cycles, then cease operating. This is because an inductive load such as the coil of a relay can produce very large voltage spikes (hundreds or even thousands of volts) if it is abruptly turned off. Consider the circuit of Figure 5-10a: when the relay coil is energized, current will flow through it. When the coil is switched off, because the relay coil is an inductor, the current will keep trying to flow. The coil will raise the voltage at the collector of the transistor value to try to keep the current flowing. If the voltage at this point happens to be greater than the breakdown voltage rating of the transistor, this tends to damage the transistor. In many cases, this circuit may operate successfully a few times and then fail after the transistor becomes sufficiently damaged by multiple voltage spikes.

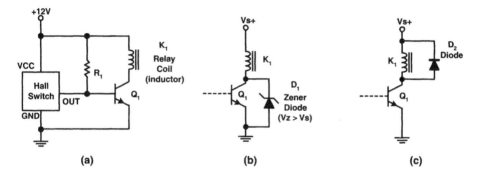

Figure 5-10: Driving a relay load. In a way that may damage the transistor (a). Protecting the output transistor with a zener diode (b) and a rectifier diode (c).

Having a circuit fail after a few or even a single operating cycle is clearly unacceptable in the overwhelming majority of applications one could imagine. There are, however, several simple techniques that can be used to protect the output transistor from an inductive load. One way is to connect a zener diode (D_1) at the transistor's collector, as shown in Figure 5-10b. While this works, it requires that the zener diode be able to dissipate the bulk of the energy stored in the inductor. As an example, for a relay operating at 12V and 250 mA, you would want to select a diode with a zener voltage of >12V, let us say 16V for this example. If you pick a zener diode with a zener voltage less than the supply voltage, it will be turned on all the time, and continuously conduct current. When the relay turnes off, the diode will have to handle a current spike of 250 mA (the relay ON current), with a drop of 16V (the zener voltage), resulting in a 4W peak power dissipation. While this power dissipation may only need be sustained for a few microseconds, it is important that the diode can handle it (see the manufacturer's datasheets!). When taking this approach, it is also important to select a transistor that has a collector breakdown voltage that is higher than the zener diode's breakdown voltage, or the transistor will simply break down before the zener begins conducting. In cases where an inductive load requires high-voltage or high current, a suitable zener diode may be either difficult to find or excessively expensive.

Another protection circuit is shown in Figure 5-10c. When the coil is turned off and the collector voltage rises above the supply voltage, diode D_2 will turn on, effectively shorting out the coil. This circuit provides the advantage of dissipating most of the inductor's energy through the inductor's winding resistance. This circuit configuration is commonly called a *fly-back diode*. An additional benefit of using a fly-back diode protection circuit is that you can use ordinary diodes as opposed to zener diodes. Suitable rectifier diodes tend to be less expensive than zener diodes of comparable power and current ratings. There are three major requirements for the diode. The first is that it can handle the coil current; diodes are readily available that can handle several amperes continuously. The second requirement is that the diode's reverse breakdown voltage be greater than the supply voltage; again diodes with reverse breakdown voltages of up to several hundred volts are easy to find and inexpensive. The final requirement is that the diode be able to switch on quickly enough to keep the voltage at the collector of the output transistor from rising too high. Ideally, the fly-back diode will limit the transistor collector voltage to the supply voltage plus 0.6V (one diode drop). In reality, the actual peak voltage may be somewhat higher; how much higher is dependent on the characteristics of the coil, the driver circuit, and the fly-back diode used.

5.12 Wiring-Reduction Schemes

If you have a large number of digital Hall-effect sensors in a system, wiring and interconnection can become a serious issue from size, manufacturability, and cost standpoints. This section presents several approaches to reducing the number of wires needed to read sensor status. The trade-off, of course, is that you will need some additional electronics.

5.13 Encoding and Serialization

If only one digital Hall-effect sensor in a group is activated at a time, it is possible to have it impress an identifying code on several wires. Figure 5-11 shows one method for encoding up to 2^N-1 digital sensors on N data lines. Diodes are used to isolate the data lines from each other. If the outputs of each sensor were tied directly to several wires, they would all be shorted together, and the scheme would not work. While the output voltage on the data lines will rise to the V_{CC} voltage when HIGH, they will only go down to about 0.7V when LOW, because of the voltage drop caused by the diodes. Because the LOW voltage will be about 0.7V, some additional interface circuitry may be needed before interfacing the outputs to a microcontroller or other logic circuitry.

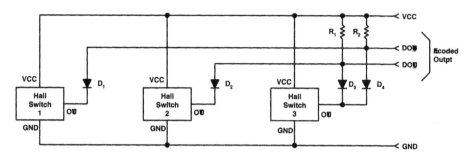

Figure 5-11: Encoding multiple digital Hall-effect sensors.

In the previous circuit, if more than one Hall-effect sensor were activated, it resulted in an ambiguous output. If the possibility exists that multiple devices can be activated simultaneously, this circuit cannot be used reliably. In cases like these, it is possible to add more electronics to be able to read devices independently of each other. An example of a system that allows independent sensor polling is shown in the block diagram of Figure 5-12.

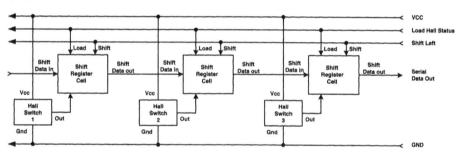

Figure 5-12: Serial shift-register multiplexing.

In this scheme, the status of all the Hall-effect sensors is first read into a string of shift register cells when the LOAD line is pulsed. Then, each time the shift line is pulsed, the status of each shift register cell is passed along to the one to its right. By loading the status of all of the sensors once, and then by repeatedly shifting the data out one bit at a time, it is possible to sequentially read the status of all of the sensors. The advantage of this technique is that it only requires five wires (VCC, GND, Shift, Load, and Serial Data Out), regardless of the number of sensors.

5.14   Digital-to-Analog Encoding

If one has a spare analog-to-digital input line with which to read sensor status, it is also possible to encode the status of multiple sensors as an analog signal. Figure 5-13 shows how the outputs of a few digital sensors can be encoded into a single binary-weighted analog signal with the addition of a few resistors.

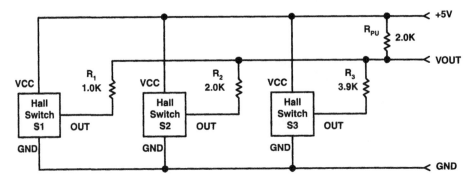

Figure 5-13: Binary-weighted encoding of digital sensors into an analog signal.

Table 5-1 shows the relationship between the output states of each sensor, and the resulting voltage at V_o. In this example the resistors are chosen to increase by a factor of two (binary weighting) between adjacent devices, ensuring that each combination is unique. One feature of this technique is that it is possible to independently monitor the status of any sensor in the circuit.

The voltages seen in any given implementation will be slightly different, varying as a function of resistor tolerance and because of the effects of the output saturation voltage of the Hall sensors used. Because of these resistor tolerance and V_{SAT} error issues, the number of sensors which can be encoded together in this manner will be restricted, with four or five likely representing a practical upper limit.

Table 5-1: Output voltage vs. sensor state for circuit of Figure 5-13.

Sensor Status			Output Voltage (V_o)
S1	S2	S3	
OFF	OFF	OFF	5.00
OFF	OFF	ON	3.31
OFF	ON	OFF	2.50
OFF	ON	ON	1.99
ON	OFF	OFF	1.67
ON	OFF	ON	1.42
ON	ON	OFF	1.25
ON	ON	ON	1.11

Another digital-to-analog encoding scheme can be used when one only needs to know the identity of the first device activated in a string of digital-output sensors. This circuit is shown in Figure 5-14. One application in which this circuit could be used is in liquid level measurement. A float containing a magnet is set up so that it moves past a line of sensors, its exact position dependent on the level of some fluid, such as water. The rightmost sensor actuated represents the level of the float, and consequently the level of the liquid. Because the rightmost sensor actuated shorts the resistive chain to ground, it is irrelevant if any sensors to the left are also activated; the output voltage is determined solely by the position of the rightmost sensor.

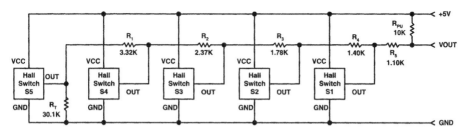

Figure 5-14: Encoding the rightmost sensor of a string as a voltage.

Note that in this circuit, the values of the resistors are not equal; they were chosen so that the output voltage would decrease in roughly equal steps. Equal-valued resistors would also result in a monotonically decreasing output voltage, but with different step sizes between output voltage levels. Table 5-2 shows the output voltage as a function of rightmost activated sensor. As in the prior analog-to-digital encoding example, we are ignoring the effects of resistor tolerance and sensor output saturation voltage. Nevertheless, with proper resistor selection it is possible to chain a significant number of sensors into this type of a "thermometer" sensing arrangement before it becomes difficult to reliably identify which sensor in the chain is activated.

Table 5-2: Output voltage vs. sensor activated in Figure 5-15.

Rightmost Sensor Activated	Output Voltage V_O
S1	0.50
S2	1.00
S3	1.50
S4	2.00
S5	2.50
None	4.00

5.15   Mini-Networks

Another way to reduce wiring is to network the sensors together by pairing each sensor with a microcontroller. In the past few years, both the cost and package sizes of microcontrollers have fallen dramatically. Microcontrollers are now available for less than $1 US in moderate quantities and are available in SOIC-8 and smaller packages. As an example, Microchip's PIC10F200 provides an 8-bit processor core, 256 words of user-programmable ROM, and 16 bytes of RAM in a 6-pin SOT-23 package, making it actually smaller than many contemporary Hall-effect sensors. The continuing downward trends in both size and cost make microcontrollers increasingly attractive options for adding inexpensive intelligence to one's designs.

Figure 5-15 shows the general organization of a simple sensor network. Each Hall-effect sensor is associated with a microcontroller that can monitor the sensor's output and provide an interface to a common data bus. Each sensor-microcontroller combination is often referred to as a network node. The network may be controlled by a central "bus master," which coordinates data transfers or, alternatively, each sensor node may have the ability to initiate and coordinate data transfers on its own—this latter arrangement is often called a *multi-master* system.

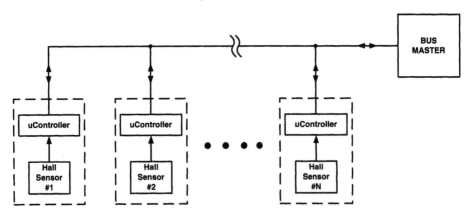

Figure 5-15: Simple sensor network.

When one thinks of a sensor network, the first options that may come to mind are Ethernet or industrial sensor busses such as Profibus or Fieldbus. While these networks have their advantages, they can require a substantial amount of processing power and software to be effectively realized. For the purposes of wiring reduction, much simpler protocols can often be used, especially if data transmission rates are relatively low and not over long physical distances. Examples of where such a network might be used are in a high-performance copying machine or automobile. Both systems must monitor numerous sensors and do so economically.

There are many options available for inexpensive network protocols. Some of the major factors in selecting a network protocol are the number of sensors that need to be monitored, the physical separation of sensors, and the speed (bandwidth). Some options available to the designer are:

SPI – Developed by Motorola, this simple-to-implement synchronous serial bus offers bandwidths up to several megabits/second over short distances, and is primarily intended to facilitate communications between chips on the same circuit card. Many low-end microcontrollers have built-in hardware to support this standard, but at lower speeds (a few tens of kilobits/second) SPI functionality can often be implemented in software.

I²C – Developed by Philips as a means of allowing chips on the same board to communicate, I²C provides bandwidths up to around 400 kbits/sec. I²C provides more functionality than SPI, and many low-end microcontrollers also provide support hardware on-chip.

CAN bus – A relatively sophisticated bus developed by Bosch for automotive applications. It offers high bandwidths and many advanced features that make it suitable for a wide range of applications. A few microcontrollers are now available that offer CAN bus support, but at this time they still tend to be relatively expensive when compared to Hall-effect sensor ICs.

Linbus – This bus was recently introduced as a low-cost, low-bandwidth (20 kilobits/second) way of monitoring sensors and controlling actuators in automobiles. A few microcontrollers are starting to come on the market with dedicated hardware support for Linbus. It is also possible to implement Linbus protocols completely in software with a small amount of external interfacing circuitry. This is definitely a standard to watch, as its increasing acceptance in the automotive industry will result in lots of inexpensive compatible chips and support.

Note that using a microcontroller as the heart of a sensor network node can provide many capabilities. First, if an analog-to-digital converter is available on the microcontroller, the network node can monitor the output of a linear-output sensor in addition to digital output devices. Because microcontrollers typically provide several I/O pins, it may also be possible for a single node to monitor several sensors—for example, a linear Hall-effect sensor may be used as a position sensor while a thermistor provides a temperature reading. Finally, the microcontroller's processing power can be used for local data reduction, reducing the amount of bandwidth needed for data transmission as well as the amount of processing power needed by a central monitor.

5.16   Voltage Regulation and Power Management

While most digital-output Hall-effect sensors contain internal voltage regulators that allow them to operate over an extended range of power supply voltages, most linear-output sensors do not, and require a regulated power supply for proper operation. Additionally, if you plan on using microcontrollers, discrete logic, or many other kinds of support circuitry, you may find that you need a source of regulated power internal to your sensor assembly, to power these other devices.

There are two fundamental approaches to providing a regulated power supply; design it from discrete components, or buy an integrated regulator in the form of an off-the-shelf integrated circuit. Unless you have a really unusual set of design requirements, the do-it-yourself approach is not recommended, as it takes a lot of discrete circuitry and design acumen to provide performance comparable to that of even the least expensive integrated regulator. For this reason we will focus on the characteristics and use of integrated devices.

Among the most popular voltage regulators are the members of the LM78xx family of linear regulators. "Linear" in the context of a voltage regulator means that the device does not contain any switching elements, and operates in a continuous manner, with no oscillators or clocks. One especially useful type of linear regulators is the 78L05 and its related variants, which provide a regulated 5V output voltage with a peak current output of 100 mA (for many versions). This device is available in both TO-92 3-leaded transistor and SOIC-8 surface-mountable packages. To use one of these parts, you provide an input voltage ranging from 8–24V to one pin, tie another pin to system ground, and 5V ($\pm$ a specified tolerance) appears on the third pin. To reduce levels of noise and provide some power-supply decoupling, it is also advisable to connect small capacitors between the input from both the inputs and outputs to ground, as shown in Figure 5-16.

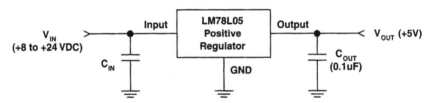

Figure 5-16: LM78L05 5V linear voltage regulator circuit.

Although these devices are about as easy to use as was just described, they have certain characteristics that one must understand in order to employ them effectively. The first is that they have definite limits on the amount of power they can safely dissipate. If one draws 100 mA from the 5V output of one of these devices while running from a 24V input, the regulator will dissipate nearly 2W of power, in the form of heat. The power dissipation for a linear regulator is approximately equal to $(V_{IN} - V_{OUT}) \times I_{OUT}$, neglecting a small amount of power consumed for internal "housekeeping" functions. This means that large differences in the input and output voltages will severely limit the amount of current one can draw from the device. For regulators in smaller packages (TO-92, SOIC-8) it can be easy to exceed the device's maximum power ratings. Maximum power ratings for integrated regulators also decrease with increasing ambient temperature. Fortunately, the manufacturers of these devices will usually provide information on derating the component for operation at higher temperatures.

The second important characteristic of these devices is that there is a "drop-out" voltage for the input below which regulation is lost, and the output is not guaranteed to be 5V (for a 5V regulator). For 78Lxx devices, this dropout voltage is usually specified as 2–3V above the output voltage; a 5V regulator would begin to "drop out" at around 7–8V of input voltage. When operation from lower power supplies is required, one can often use what is known as a low-drop-out regulator often referred to simply as an LDO. One family of such devices is the TL750 series produced by Texas Instruments. These devices can maintain regulation when operating with dropout voltages of less than a volt. One important consideration when using many types of LDOs is that the capacitive load on the device's output is a critical factor in maintaining stability. Most LDOs have very specific requirements for the amount of capacitance you bypass their outputs with, and even the type of capacitor you use! If you put either too much or too little, or even the wrong kind of capacitor on a given device's output, it may begin oscillating. Fortunately, manufacturers of LDOs tend to provide very specific information about acceptable values and types of bypassing capacitors in their devices' datasheets.

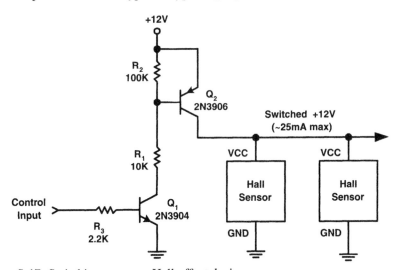

Figure 5-17: Switching power to Hall-effect devices.

Another power-related issue is that of power management. Because most Hall-effect sensors consume several milliamperes of operating current, they can be difficult to use in battery-operated systems that must operate for an extended length of time. One solution to this problem is to shut them off when not in use. The circuit of Figure 5-17 provides one way to do this. When the control input is at 0V, transistor Q_1 and, consequently, transistor Q_2 are both off. When 5V is applied to the control input, both transistors are turned on. Because Q_2 is a PNP device, it can go into saturation, and introduce a relatively small (<100 mV) voltage drop when sourcing moderate amounts of current (a few tens of milliamperes in this case). While this small voltage drop should pose little problem when power-switching digital Hall-effect sensors in this manner, it should be considered when attempting to use this technique to power-switch a ratiometric linear-output device, where the small drop in supply voltage will be reflected as changes in zero-flux offset and sensitivity.

Chapter 6

Proximity-Sensing Techniques

Sensing an object's presence or position are two of the most widespread applications in which Hall-effect sensors are used. Magnetic field sensors are well suited for this kind of application for two reasons. The first is that magnetic fields are not significantly affected by nonmagnetic materials and pass through them unhindered. The second reason is that strong magnetic fields do not occur often in nature (at least on Earth) and if a strong magnetic field is encountered, it is usually manmade. Strong magnetic "interference" sources are not very common. This rarity of accidental sources of magnetic interference makes a strong magnetic field a good indicator. While most objects one might want to detect won't produce significant magnetic fields, it is usually a simple enough matter to affix a small permanent magnet to them to provide a readily detectable field.

This chapter will describe some of the ways in which permanent magnets and Hall-effect sensors can be used to detect proximity and measure position of objects.

6.1 Head-On Sensing

When people begin working with Hall-effect sensor ICs, one of the first things they usually do is to take a permanent magnet and bring one of the poles up to the device to activate it—to "try it out." This operation mode is called *head-on* actuation, and is shown in Figure 6-1a. Head-on actuation is one of the most common methods for activating Hall sensors in binary (on-off) proximity detection applications.

In a head-on application, the sensitive axis of the sensor and the axis of magnetization are co-linear. The magnetic flux-density that the Hall-effect sensor sees as a result of being approached by a magnet pole is highly nonlinear with respect to magnet-sensor airgap. It decreases rapidly as airgap increases, as shown in Figure 6-1b. A curve like this, relating magnetic flux to physical position, is called a *flux map* or a *magnet map*. Such maps are an extremely valuable tool for developing Hall-effect-based applications, and will be used throughout this and subsequent chapters.

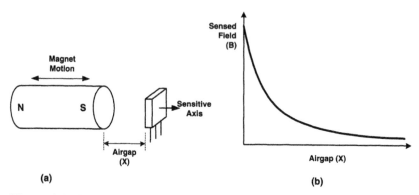

Figure 6-1: Head-on actuation mode (a) and flux density vs. airgap (b).

To get a present/absent binary output, switch-type digital Hall sensors are used in a head-on sensing system. As one approaches the magnet, the field increases as an inverse function of distance. When the field exceeds the operate point (B_{OP}) of the sensor, it will activate. As one moves away from the magnet, the sensor will deactivate when the field drops below its release point (B_{RP}). Because the magnet's field along its axis of magnetization always remains the same polarity regardless of the distance traveled along that axis, a switch-type device must be used if one wants the sensor to turn OFF. Figure 6-2 shows how mechanical operate and release points can be determined from superimposing the magnetic B_{OP} and B_{RP} points on a flux map.

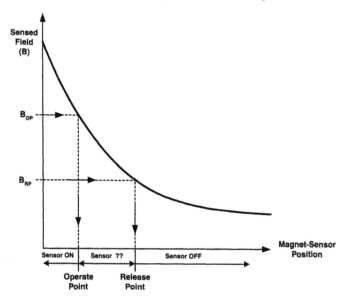

Figure 6-2: Determining mechanical operate and release point from B_{OP}, B_{RP} and a flux map.

While it is possible to get consistent mechanical operate points by using a sensor with a high B_{OP} switch point, so that it turns ON close to the face of the magnet, release points tend to be less consistent, especially for parts with large hysteresis values (B_H). This is because the gradient of field vs. distance is greater near the magnet than further away. This means that the closer you get to the magnet, the more sensitive the field measurement will be as a function of position. The nonlinear character of the magnetic field vs. distance also makes it difficult to design a magnet-sensor combination that meets an arbitrary specification for mechanical turn-on and turn-off points. For these reasons, head-on sensing is often best employed in applications where tight position-sensing tolerances are not necessary.

One can also use a linear-output sensor in an effort to measure the actual magnet-to-sensor distance. The nonlinear flux vs. airgap characteristics of head-on make it less than optimal for most linear position measurement applications. Later in this chapter we will discuss magnetic systems that are more suited for sensing linear position.

6.2 Slide-By Sensing

Another common way of using a Hall-effect sensor as a proximity detector is by sliding the pole-face of a magnet past the device, as shown in Figure 6-3a. In this scenario, the magnet's axis of magnetization and the sensitive axis of the Hall sensor are both parallel, but the magnet moves perpendicularly to the axis of magnetization. This sensing method is particularly useful when there is a significant chance that the magnet will travel past its normal end-stop position. In the case of a head-on sensing configuration, such over-travel could damage either the sensor or the magnet.

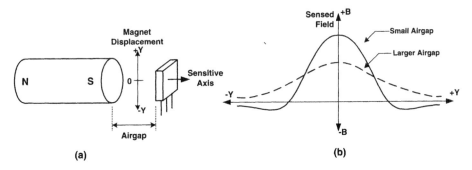

Figure 6-3: Slide-by actuation mode (a) and resultant flux vs. position (b).

Because the magnet can travel past the sensor, the slide-by configuration is only useful for indicating when the magnet is actually in front of the sensor; if one uses this configuration for an end-of-travel indicator, then a separate mechanical stop must be provided to limit the magnet movement.

In situations where it can be used, the slide-by configuration provides some significant advantages over the head-on configuration. The first is that, as the magnet moves from the center ($x = 0$) position, the sensed magnetic field may eventually drop slightly negative, as shown in Figure 6-3b. This occurs because when the sensor is off center past the edge of the magnet, it will be sensing the field returning back to the rear pole. This provides a guaranteed turn-off for any kind of switch-type ($B_{RP}{>}0$) Hall IC. This effect will be particularly pronounced for small sensor-to-magnet airgaps. The physical displacement at which the field crosses zero is also fairly consistent with respect to small variations in the effective gap between the magnet and sensor. This means that by using a Hall switch with low B_{OP}/B_{RP} points, consistent physical operate and release points can be obtained. This also means that the width of the ON region can be controlled by the width of the magnet employed. In many cases, for a switch with sufficiently low B_{OP} and B_{RP} points, the width of the on-state region will closely match the width of the magnet pole-face, even over significant variations in magnet-to-sensor spacing. Because the negative field past the magnet edges may be small in relation to the positive field at the magnet face, a switch-type sensor will often be the best choice of sensor type because the negative field eventually tapers off to zero at some distance, and even at the magnet edge may be insufficient to provide a guaranteed turn-off for a latch-type sensor.

6.3 Magnet Null-Point Sensing

While the last sensing methods relied on detecting some finite positive field to trip a sensor, the next class of sensor-magnet configurations to be discussed rely on detecting null-points in the field around a magnet, or places where the net field in a particular axis is zero. These techniques are especially useful in situations where a high degree of switching accuracy over a small amount of total travel is needed. Using positive flux to denote the ON region and negative flux to denote the OFF region provides several advantages:

- The positions of magnetic null points tend to be stable over temperature, especially for magnetic systems consisting of a single magnet of a single material.
- Highly sensitive symmetric latch-type Hall ICs can be used. These parts, particularly auto-nulling types, can be very stable over variations in temperature and power supply conditions.
- It is possible to create very sharp transitions between negative and positive fields with the proper magnetic circuits, allowing for fine control of actuation points. The rate of this transition is referred to as the *gradient*.

Figure 6-4 shows two methods of developing a magnetic null point from a single magnet. The arrow emanating from the sensor indicates the sensitive axis. Both of these techniques are useful in slide-by type applications.

The configuration of Figure 6-4a detects the normal (perpendicular) flux emanating from one of the nonpole sides of a rectangular bar magnet. Halfway between the poles, this flux is zero, and gets stronger as one approaches either pole (this is why steel objects aren't attracted to the middle of bar magnets, but to the ends).

Figure 6-4b shows a sensor oriented to detect the flux parallel to a magnet pole face. Because the flux diverges from the pole, its pole-surface-parallel component is negative to the left of centerline, positive to the right of the pole center, and zero at the centerline.

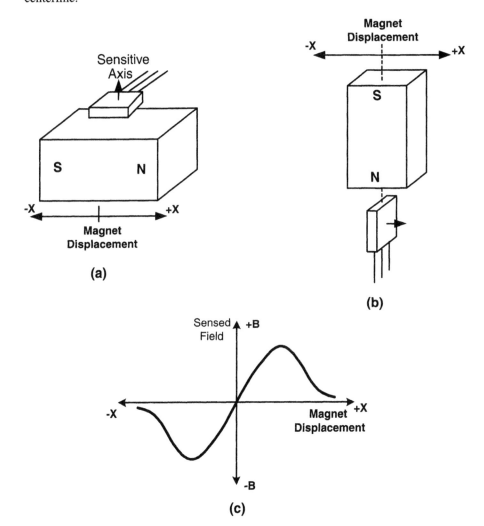

Figure 6-4: Methods of creating magnetic null-points.

Although both of these configurations will provide roughly the response shown in Figure 6-4c, there can be significant differences both in the shape and magnitude of the response, resulting from measuring the field at the pole face versus measuring the field along the length of the magnet. Because the packaging of most Hall-effect sensors allows one to position the transducer element closer to the magnet surface in the example presented in Figure 6-4a, this configuration will usually provide the highest magnetic gradients, and the sharpest switching points of the two alternatives.

In general, when using comparable magnets, the configuration of Figure 6-4a will tend to provide both a greater response and steeper gradients than the configuration of Figure 6-4b.

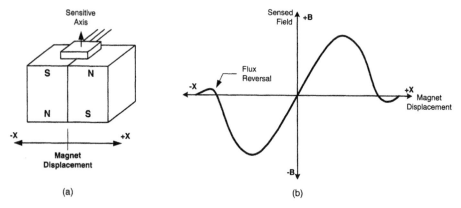

(a)   (b)

Figure 6-5: Developing high magnetic gradients with a compound magnet.

To develop even sharper gradients, the system shown in Figure 6-5 may be used. By placing two magnets alongside each other in an antiparallel configuration, a very sharp transition in the flux normal to the pole-faces can be made to occur, because of the abrupt transition between the north and south poles. One disadvantage of this approach, however, is that fringing effects cause flux reversals past the pole faces, as previously discussed in the case of the slide-by configuration. If the magnetic sensor is either sufficiently sensitive, or sufficiently close to the actuator magnet pair, false actuation can occur when using a compound magnet.

In all of the above cases, it should be noted that if one uses a latch-type digital Hall-effect sensor, the output will only be valid when the sensor is actually close enough to the magnet to be positively switched either ON or OFF. Because a switch-type sensor will be OFF in the absence of magnetic field, it provides an indication that the magnet is not present. If a latch-type sensor, however, is moved far enough from the magnet so that it is no longer affected by it, it will simply maintain whatever its last state was. Similarly, if the latch-type sensor is outside the influence of the magnet when it is powered up, its state is unpredictable. If one plans on using latch-type devices in a magnetic null-point sensing configuration, provisions must be made to either ensure that the magnet does not over-travel past the sensor, or that some separate indication of over-travel conditions is provided.

One example of a system in which a null-point sensor could be useful is a linear rail positioning system. In a linear rail positioning system, a bearing block travels back and forth along polished steel rods, often propelled by either a lead screw or a drive belt. In many of these systems, it is important to know when the block has reached its end of travel, primarily so that the system has an absolute position reference (home position), and secondarily so that the system is not mechanically damaged by over-travel. An example of how a null-point sensor might be used in such a system is shown in Figure 6-6.

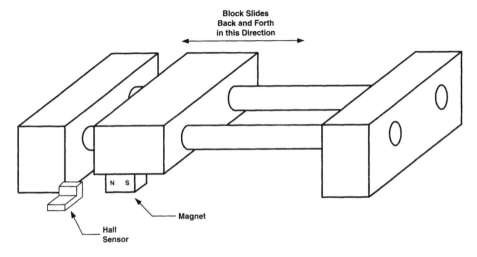

Figure 6-6: Linear rail guide system with Hall-effect home-position sensing.

Because the state of the sensor should always be HIGH when the bearing block is anywhere between its physical end-of-travel position and the home position, one needs to position the magnet and sensor so that they provide a response such as that shown in Figure 6-7.

In this system, "end-of-travel" is indicated by a ON output from the Hall-effect sensor. Because it is important to be sure when the bearing block is near the end-stop or not, a sensitive switch should be used instead of a latch. This will ensure that the "near-home" status (ON) will only be reported when the bearing block is actually near the end-stop. The home position itself is determined as where the sensor makes the OFF-ON transition. Because the magnetic hysteresis of the sensor will translate into hysteresis in the mechanical measurement, the home position will therefore be slightly different depending on the direction in which it is encountered. For this reason the

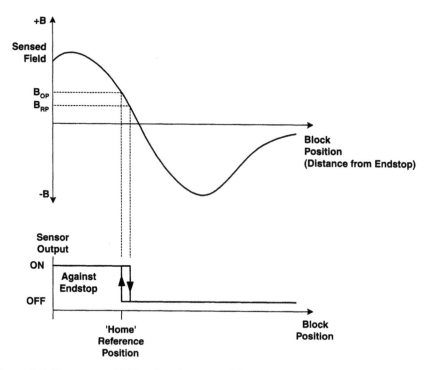

Figure 6-7: Response of Hall-effect home-position sensor.

home position must always be determined when moving in the same direction (i.e., to-wards the end-stop). With selection of an appropriate magnet and sensor, home position repeatability and mechanical hysteresis on the order of a few-thousandths of an inch are possible. This performance would be difficult, if not impossible, to achieve with the simpler head-on and slide-by configurations.

6.4 Float-Level Sensing

One position-sensing application that uses yet another magnetic configuration is float-level sensing. In this application, the sensor assembly is typically used to measure the level of a liquid or to determine whether that level exceeds some given low-point or high-point. Figure 6-8 shows a schematic view of a Hall-effect-based float-level sensor. A donut-shaped ring magnet is used to provide the actuating field. This magnet surrounds a shaft and is free to both move along the stem and may also be allowed to rotate. The magnet is embedded in a float assembly, which can be a hollow ball, or a piece of buoyant material such as closed-cell plastic foam. Inside the shaft are one or more Hall-effect sensors, which are used to detect the position of the magnet.

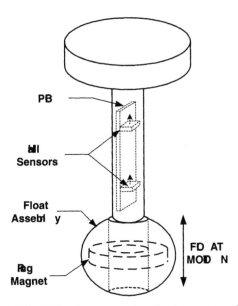

Figure 6-8: Example of liquid-level sensor using float magnet and Hall-effect sensors.

In this magnetic configuration, the ring magnet is magnetized parallel to its major axis (through the donut-hole). If the inner diameter of the ring magnet is of comparable dimension to its height, there will be a significant field inside the inner diameter, with a polarity opposite to that in which the ring magnet is magnetized. The sensors in the stem are oriented so that they detect the field aligned along the direction of the stem. Figure 6-9 shows a cross section of the field surrounding a ring magnet, and also a graph of the field along the major axis.

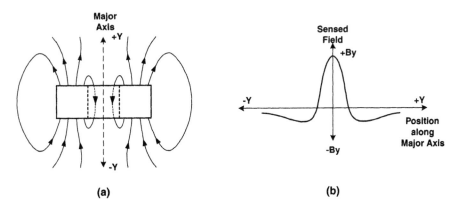

Figure 6-9: Cross section and graph of flux-density along major axis of axially magnetized ring magnet.

When designing a float-level sensor, there are a few points worth keeping in mind. First, if one is trying to make a minimum or maximum level switch, one must put a mechanical end-stop at the appropriate points to make sure that the float doesn't travel past the sensor. If one is trying to determine actual level, as opposed to merely making a less-than/greater-than type of measurement, one must use multiple sensors spaced along the length of the stem. In this case it is important to make sure that the sensors are spaced closely enough together so that the magnet actuates at least one of them at all times; otherwise you will have measurement "dead-spots." Finally, because the field inside the ring magnet's inner diameter does not present any steep gradients, it can be difficult to make a float sensor with highly accurate trip points (unit-to-unit) when using this magnetic configuration. Low trip-point accuracy can be worsened by the use of floats that fit loosely on the stem and are allowed to tip into an off-axis orientation. Despite potential issues of low trip-point accuracy, this configuration provides a simple and robust solution to the problem of measuring liquid level.

6.5 Linear Position Sensing

In addition to being useful as binary presence/absence detectors, Hall-effect sensors can be used to measure continuous displacement. To perform this measurement requires two items: a linear-output Hall transducer, and a magnetic circuit that provides a magnetic field that varies monotonically as a function of displacement over some specified range of motion. While a linear magnetic response is not essential, as "correction" can be applied later, either electronically or in software, a linear response does make the system easier to work with. If the magnetic response versus displacement is sufficiently linear, no correction may be needed in many applications.

Simple continuous position sensors can be implemented by taking any of the proximity-sensing schemes described above and replacing the digital sensor with a linear output device. While such implementations may be useful for particular applications, they suffer because the output will be significantly nonlinear with respect to position. Another problem encountered will be sensitivity to mechanical tolerance between the magnet and the sensor.

Fortunately, there are several magnetic configurations that can provide near-linear flux density as a function of mechanical displacement over distances comparable to the size of the magnets employed. One such scheme employs two magnets held a fixed distance apart by a nonmagnetic yoke with similar poles facing, as shown in Figure 6-10. The flux density in the X direction between the two magnets varies from negative minimum at one pole face, reaches zero halfway between the poles, and increases to a positive maximum at the pole face of the other magnet. Near the halfway point, the slope is nearly constant, resulting in a region with an approximately linear transfer function. One feature of this arrangement is that it is possible to select magnets that will provide a linearly varying field over most of the distance between them. If the dimensions of the pole faces are large compared to the pole separation, the field as seen by the Hall-effect sensor will also be relatively insensitive to translations in sensor position in directions other than the X axis.

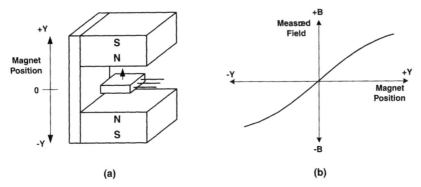

Figure 6-10: Magnets arranged to give linear varying field in gap.

Although this arrangement provides a magnetic flux density proportional to displacement, with a zero-flux null point in the middle, it doesn't allow for the possibility of overtravel. When one cannot guarantee that the magnets won't stop moving before they hit the sensor, the magnet configuration of Figure 6-11 can provide an alternative solution.

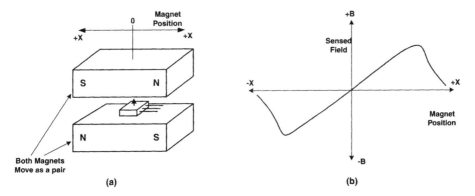

Figure 6-11: Another linear flux density vs. position scheme.

The advantage of this latter arrangement is that the sensor won't crash into the magnets under an over-travel condition. With the appropriate selection of magnet materials and geometries, it also provides a linear flux variation in response to motion. The primary drawback is that, as the sensor moves out of the gap in the direction of travel, the measured flux begins to drop, which can cause false position readings. For this reason, one still needs to consider the effects of over-travel, but from a system performance standpoint.

When considering the use of one of the above schemes, there are several issues to keep in mind. The first is that the length-of-stroke over which measurements can be made will be comparable to the size of the magnets. Unless one wants to use very large magnets, one will be limited to small ranges of mechanical measurement.

The second issue is in choice of magnet materials. While rare-earth magnet materials such as NdFeB or SmCo are often a logical choice for actuators in a binary (On/Off) proximity detection system because of their high field strength, they may not be as efficacious in a linear position-sensing system. This is because many linear sensor ICs are designed to operate over the range of ±1000 gauss or less; the 3000–5000 gauss obtainable from a rare-earth actuator can saturate the sensor, shortening the effective measurement range to a small fraction of the total mechanical travel. In cases where linearity over a wide travel range is required, the use of Alnico or ceramic magnets should also be considered, as their flux output in many configurations is comparable to the sensing range of many linear Hall-effect sensors.

While we have described the magnetics required to obtain linearity, we have ignored issues such as offset and span. Offset errors will result from assembly tolerances and from inhomogeneities in the magnets. Span errors will result from variations in magnetization, as well as from tolerancing errors. As long as the resulting system (magnetics+sensor) is linear, however, offset and span can often be accounted for and trimmed out downstream from the sensor in subsequent signal processing.

6.6 Rotary Position Sensing

Sometimes one needs to know the angular as opposed to the linear position of an object. While it is possible to create mechanical linkages that convert a rotary motion into a linear displacement, there are also ways to measure rotary motion directly. Figure 6-12 shows one of the simplest methods, which is to rotate a uniform field around a Hall-effect sensor.

If the Hall-effect sensor is placed at the center of rotation, it will always be exposed to the same field, but from a different direction. Because a Hall-effect sensor only is responsive to field components in a single axis, this results in a response that is of the form $V_0 = k \sin \theta$. While not linear, this function is monotonic over a range of ±90° of rotation. This means that it is possible to determine the angle from the sensor's reading over this range through application of the $\sin^{-1}$ function. For practical sensors, the necessary conversions can often be implemented as look-up tables in a microcontroller.

Gain and offset adjustment are critical operations that must be performed on the transducer signal before it is passed through a $\sin^{-1}$ correction (Figure 6-13). Offset needs to be adjusted when the assembly is in the 0° position. The net flux at this point, as seen by the sensor, should be very close to zero, assuming no asymmetries in the magnets or their locations. The dominant source of offset error at 0° will often be the transducer. When the magnets are moved to 90° (or the sensor assembly's maximum functional span) gain adjustment can then be performed. The gain errors will be a combination of flux variation in the magnet (degree of magnetization) and the transducer gain.

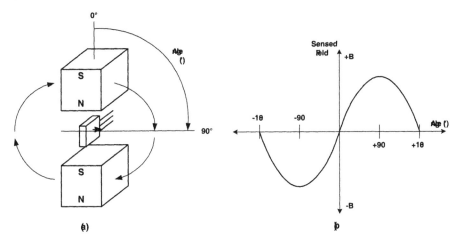

Figure 6-12: Rotary position sensor and response.

Figure 6-13: Signal correction for rotary sensor.

If a smaller angular sensing range is required, it may be possible to ignore the nonlinearity of the sinusoidal response. Response over the range of ±30° or less often is sufficiently close to linear to be useful in many applications without requiring any sin⁻¹ correction. In this case it is also desirable to use higher levels of flux in the magnetic circuit so as to increase sensitivity.

The addition of a circular flux concentrator (Figure 6-14) can be a useful improvement to a rotary position sensor's magnetic circuit. By routing the flux through a closed magnetic circuit, the concentrator greatly reduces the flux that can escape into the outside environment and possibly interfere with any other magnetic devices. Through the same shunting mechanism, the concentrator may also reduce the measurement errors caused by external magnetic fields. Finally, by shortening the overall path the field must take though the magnetic circuit, a concentrator can allow for the use of less magnet material, or lower grades of magnet material, reducing overall system cost.

What if angular measurements are needed over more than ±90° of rotary travel? One solution is to use two Hall sensors, placed at right angles to each other, as shown in Figure 6-15. In this configuration one obtains outputs that are proportional to sin θ

and cos θ functions. By looking at the relative polarity and magnitude of the two output
signals, angle can be determined over the entire 360° range.

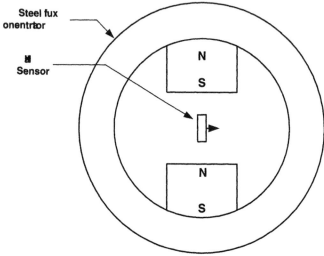

Figure 6-14: Circular flux concentrator.

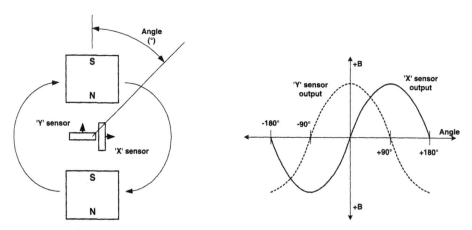

Figure 6-15: Sensor for 360° (±180°) operating range.

While one can resolve angle by processing the outputs of each sensor separately,
with sin⁻¹ and cos⁻¹ correction, and determining the quadrant of operation by considering the signs of the individual outputs, there is a better way to perform this function.
The angle (θ) can also be derived by considering the function:

$$\theta = \tan^{-1}\left(\frac{B_Y}{B_X}\right)$$ (Equation 6-1)

Deriving angle as a function of the ratios of the two sensor outputs provides several advantages. The first is that one need only match the gains of the two transducers. As long as the magnetic circuit provides the transducers with enough flux to make accurate measurements without driving them into saturation, the exact amount of flux provided by the magnet becomes less important. For the same reason, the temperature coefficient of the magnets also becomes insignificant. Finally, if one can match the two transducer's temperature coefficients of sensitivity, transducer temperature coefficients also become much less significant. Note, however, that transducer offset error is still a consideration and needs to be minimized.

The price to be paid, however, is significantly more complex logic required to resolve the angle from the individual sensor measurements. Because Eqn. 6-1 provides a unique angle only over the span of ±90° ($B_X > 0$), a number of variations must be used to resolve angles in other quadrants. Additionally, because the ratio B_Y/B_X becomes very large as one approaches either +90° or –90°, it is better to take the $\tan^{-1}$ of the reciprocal and add an offset when $|B_Y| > |B_X|$. To accurately interpret the raw B_X and B_Y measurements requires three tests:

1) Is $B_X > 0$?
2) Is $B_Y > 0$?
3) Is $|B_y| > |B_x|$?

Based on these three criteria, one can then select an appropriate formula for resolving the angle (as measured clockwise from the positive Y as shown in Figure 6-15) from Table 6-1.

Table 6-1: Formulae for resolving 360° resolution.

Range (°)	IBxI > IByI	Bx > 0	By > 0	Formula for θ (°)
0–45	No	Yes	Yes	$\theta = \tan^{-1}(B_x/B_y)$
45–90	Yes	Yes	Yes	$\theta = 90° - \tan^{-1}(B_y/B_x)$
90–135	Yes	Yes	No	$\theta = 90° - \tan^{-1}(B_y/B_x)$
135–180	No	Yes	No	$\theta = 180° + \tan^{-1}(B_x/B_y)$
180–225	No	No	No	$\theta = 180° + \tan^{-1}(B_x/B_y)$
225–270	Yes	No	No	$\theta = 270° - \tan^{-1}(B_y/B_x)$
270–315	Yes	No	Yes	$\theta = 270° - \tan^{-1}(B_y/B_x)$
315–260	No	No	Yes	$\theta = 360° + \tan^{-1}(B_x/B_y)$

6.7 Vane Switches

Sometimes it is undesirable to affix a magnet to a moving member to sense position. Another proximity sensor that can be made with Hall-effect sensors is the vane switch. This type of sensor detects the presence or absence of a ferrous flag (the vane). In its simplest form, a vane switch consists of a Hall-effect switch and a magnet in close proximity. When the flag is not present, the Hall sensor detects the magnet's field and remains ON (Figure 6-16a). When the flag passes between the magnets and the Hall switch, it interrupts the field and the switch turns OFF (Figure 6-16b). Note that the flag doesn't block the magnetic field; it merely provides a shorter path back to the far pole of the magnet, and by doing so shields the sensor from the magnet's field.

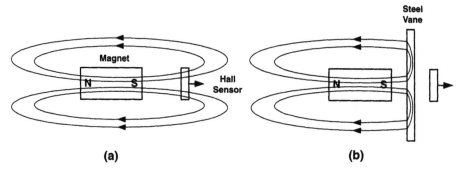

Figure 6-16: Vane switch operation. Vane absent (a) and vane present (b).

Figure 6-17 shows the general characteristics of how the magnetic flux density as seen by the sensor varies in response to the vane entering the gap between the magnet and sensor. When the vane is clear of the gap, the sensor detects a maximal amount of on-state flux. As the vane begins to enter the gap, the flux begins to drop off rapidly. Both the point at which the drop-off begins and the rate of drop-off are dependent on a combination of the magnetic design of the vane switch and on the geometry and composition of the vane flag. For practical vane switches, flux decay may begin to occur when the vane is still a significant distance from the vane switch's centerline.

When the vane has completely entered the gap, the sensor will still see a small amount of leakage flux. This is field that has passed through the vane to the sensor, as opposed to having been shunted back to the magnet. Finally, to make matters even more complex, the position of the vane within the gap will also influence the behavior.

In order to develop an effective vane switch, one must consider the following factors:
- On-state field
- Leakage field
- Mechanical start and end points of flux density roll-off
- Rate of flux density roll-off (slope)
- Matching the magnetic design to available Hall-effect devices

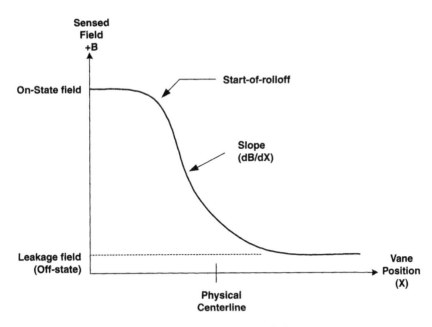

Figure 6-17: Vane switch flux map and key characteristics.

On-state flux field is important because if there isn't enough flux to activate the Hall-effect switch with no vane in the gap, the vane switch will never indicate vane absence. On-state flux density therefore needs to be greater than the sensor's maximum B_{OP} under worst-case conditions. Additionally, if B_{OP} is only slightly less than the on-state flux density, the vane switch assembly may not switch ON until the vane is a significant distance away from the assembly's centerline, possibly remaining OFF even when the vane is mechanically out of the assembly.

Leakage field presents the opposite problem. If the vane and magnetics allow too much leakage flux to reach the sensor, the Hall-effect switch may never turn OFF, resulting in a failure to indicate vane presence. The leakage flux should therefore be less than the sensor's minimum B_{RP} under worst-case conditions. In many cases this issue can also be addressed by appropriate design of the vane. Vane flags that are too small (do not shield sensor), too thin, or made of materials that saturate easily can result in excessive levels of leakage flux. For this last reason, a piece of inexpensive cold-rolled steel will often make a better vane interrupter than an expensive piece of mu-metal or permalloy (specialty high-permeability magnetic alloys), because these alloys tend to saturate more easily than common steels.

Start, end and rate of roll-off are extremely complex factors to determine a priori in a design, but should be characterized, either through finite element simulation, or measurement of prototype assemblies, in order to get some idea of how sensitive the design will be to variations in both the magnetics and the characteristics of the Hall-effect sensor.

Because adjustable-threshold (B_{OP} and/or B_{RP} are user-selectable) Hall-effect switches are, at the time of writing this book, considerably less common and more expensive than fixed-threshold devices, most economically practical vane switch designs will employ fixed-threshold devices. This means that you, as the designer, have a somewhat limited number of choices to select a Hall-effect switch from. Consequently, the magnetics must be designed to accommodate the characteristics of available Hall switches, which can make the design process quite challenging. This is especially true when one must develop a vane switch to operate and release at arbitrary points when operated with a specific target.

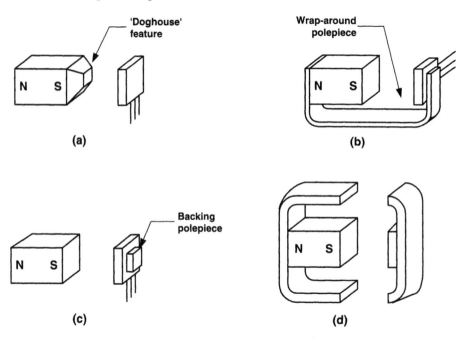

Figure 6-18: Vane switch magnetic architectures.

There are many magnetic structures that can be used to implement vane switches. Each of these structures will change both the amount of field present in the gap, and the rate at which it changes in response to an entering vane. The main goals of using complex magnetic structures in a vane sensor assembly are to reduce the cost (by allowing for the use of smaller magnets) and controlling the vane sensor's trip-points. Because of the complexity of magnetic interactions, and of designing a vane switch and vane combination for a particular application, we can't provide any specific design formulas, or even general rules for application. What follows is a general description, however, of a few potentially useful magnetic structures and what they do:

1) ***Doghouse magnet*** (Figure 6-18a). Forming the magnet into a flattened point will tend to concentrate the flux at that point. This technique is more effective with Alnico-type magnets than with rare-earth magnets, because of the lower permeability of rare-earth magnets. Using a piece of soft steel to form the doghouse tip results in even still more flux effective concentration, at least until the flux density increases to the point where it saturates the steel.

2) ***Wrap-around polepiece*** (Figure 6-18b). Wrapping a soft steel polepiece around from the back of the magnet to the back of the sensor reduces the total effective path that the magnet's flux has to travel. This intensifies the field seen by the sensor, allowing for the use of smaller magnets. It also provides the side benefits of reducing the stray fields produced by the vane assembly, and may in some cases make the vane less susceptible to influence from external stray fields.

3) ***Sensor-backing polepiece*** (Figure 6-18c). Backing the sensor with a piece of high-permeability material will intensify the field it sees, allowing for the use of a smaller magnet and providing better control over trip-points. Unlike the wrap-around polepiece, however, a sensor-backing polepiece will tend to make the vane assembly more susceptible to interference from outside fields.

4) ***E-core polepiece*** (Figure 6-18d). This represents a compromise between the wrap-around and the sensor-backing polepieces. The author used this architecture once in a design for a wide-throat vane sensor where manufacturing constraints would not allow for a wrap-around polepiece design.

It is also possible to combine various aspects of these structures to meet a given set of requirements. The development of a cost-effective vane interrupter is a nontrivial task that requires both a good knowledge of magnetics as well as a willingness to experiment.

Because Hall-effect vane switches are largely unaffected by contamination, external light, and can operate well at temperature extremes, they are often used in harsh environments as more-rugged alternatives to optical interrupters. One big difference between using an optical interrupter and using a Hall-effect vane switch is the mechanical force exerted on the vane by the magnets. The vane switch's magnet will tend to pull the vane into the airgap. For applications where significant amounts of torque are available to move the vane, such as a lead-screw end-of-travel limit switch, the sensor's mechanical drag may not be an issue. For other applications, such as a printer paper-path sensor, where the vane must be moved by a piece of paper, a vane switch may be completely unsuitable.

A commercial example of a vane-switch designed specifically to replace optical-interrupters in high-contamination environments is shown in Figure 6-19. Because it was intended as a direct replacement, both cost and housing geometry were the major design constraints. This required using a very simple magnet-sensor system, consisting of a small rare-earth magnet and a commodity bipolar switch. The result is a unit that can be used to replace optical interrupters in numerous applications by merely changing the pad-patterns of the printed circuit board into which it is to be soldered, and making sure a ferrous vane is used as a target.

Figure 6-19: Example of packaged Hall-effect vane switch assembly (courtesy of Cherry Electrical Products).

6.8 Some Thoughts on Designing Proximity Sensors

One common design trap that people fall into when developing sensor assemblies around off-the-shelf Hall-effect ICs is the deceptive simplicity of getting a prototype to work. Many designs can be tweaked so that small numbers of articles will function adequately. The problem that often then occurs is high fallout when that design is transferred to production and must function in the face of random variation in both the sensor ICs and associated magnetics. The solution to this problem is a careful consideration of the effects of component variation and tolerance.

There are three major sources of variation in a typical Hall-effect-based position sensor:

1) The magnetic parameters of the sensor IC
2) The characteristics of the magnetic materials
3) Mechanical tolerance

Hall-effect sensor IC manufacturers usually publish minimum and maximum limits for their devices' key parameters, often specified over their operating temperature ranges. These minimum and maximum datasheet limits are extremely important as design guidelines, as product may fall *anywhere* within these limits. Because of the batch nature of IC fabrication processes, devices from the same wafer or same processing lot will tend to have similar characteristics, with a statistical distribution much tighter than might be inferred from the datasheet limits. Devices from a different lot, however, while still falling within published specifications, may have a significantly different statistical distribution. Figure 6-20 illustrates this kind of lot-to-lot variation.

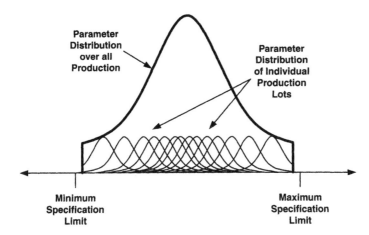

Figure 6-20: Hypothetical lot-to-lot variation in Hall-effect IC parameters.

Because of potential lot-to-lot variation, it is risky to evaluate a design solely on the basis of having a number of functional prototypes, especially if one does not know the individual characteristics of the sensor ICs used. To help ameliorate this problem, some sensor IC manufacturers can provide small quantities of characterized devices as an aid to the designer.

A similar problem arises with the magnetic components used in a design. Both permanent magnets and soft magnetic components are made in batch processes, and are subject to both variations across devices from a single batch, as well as potentially larger batch-to-batch variation. Depending on how aggressive your purchasing department is about cost reduction, you may also have to contend with variations in "identical" materials sourced from multiple manufacturers.

One serious drawback to magnetic material variation is that the majority of data-sheets for magnetic materials do not provide clearly defined (if any!) upper and lower limits for many important parameters or how they vary over temperature. The more reputable manufacturers, however, will often be willing to provide some guidance in how you should account for tolerancing key parameters if you get in touch with their applications engineering departments.

Mechanical tolerance may also be a significant source of performance variation in your sensor designs. At a minimum, you must understand how the tolerances of the various components combine or "stack up," and their performance effects, to make intelligent decisions about which tolerances need to be tightened up, and even whether the sensor has a chance of being viable in production.

While accounting for the effects of all tolerances in a sensor design can be a complex undertaking, it is possible to get a first-order estimate of some of these effects. To start with, if one knows the function describing magnetic flux vs. airgap and has the worst-case (min/max) operate and release points (B_{OP} and B_{RP}) for the sensor to be used, one can predict the range of physical switchpoints for that particular sensor-magnet combination. Figure 6-21 shows how this can be done. The first step is to place the minimum and maximum operate and release points (B_{OP} and B_{RP}) on the B axis of the flux density vs. airgap curve. Lines are then extended to the curve and dropped down to the "position" axis. The worst-case physical operate points and physical release points (P_{OPMAX}, P_{OPMIN}, P_{RPMAX}, P_{RPMIN}) can then be read off the airgap axis. Note some of the different physical regions defined by this graph. Toward the ends are regions where the sensor's behavior is well-defined (Must be on/Must be off), while in the middle there are regions where the device could be in an indeterminate (either on or off) state. Minimum and maximum B_{OP} and B_{RP} values should be selected based on the sensor assembly's intended operating temperature range.

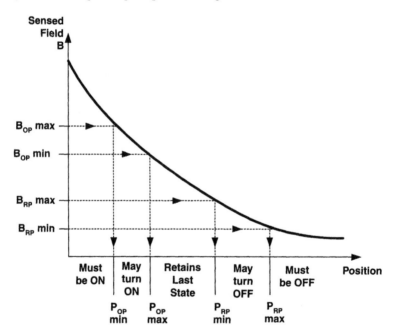

Figure 6-21: Relating B_{OP} and B_{RP} to physical operate and release points.

Once one has obtained minimum and maximum physical operate and release points, the system can also be described by a hysteresis curve relating the ON and OFF states to physical position, as shown in Figure 6-22. Again, there are several physical operating regions represented by this graph, with both well-defined and indeterminate sensor output states.

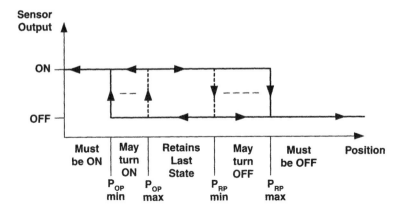

Figure 6-22: Hysteresis curve relating ON and OFF states to physical position.

If one has access to magnetic modeling software (described in Chapter 10), and can obtain minimum/maximum magnetic material models from their suppliers, it may also be possible to repeat the above exercise using a variety of magnetic flux vs. position curves. By varying magnetic material parameters and geometry in modeling software, it is possible to estimate the effects of both magnetic material and dimensional variation on the magnetic behavior of one's sensor system.

While the above method isn't a magic-bullet solution, and does not deliver magnet and IC specs based on a set of physical switchpoints, it can be useful when applied in an iterative manner, and used to move from an initial guess of magnet and IC towards a final, workable design.

Even though it is more work than simply taking the "build-it-and-see" approach, there are several tangible benefits to measuring or simulating magnet flux-vs.-airgap characteristics and iteratively performing the above exercise. The first is that you can obtain some idea of how robust your design is with respect to potential manufacturing variation. Because of the high degrees of variation in some Hall-effect devices and magnetic materials, it is quite possible to inadvertently come up with a marginal design for which the first few prototypes will work satisfactorily, but which will later experience a high percentage of fallout in production.

A secondary reason for performing an analysis such as that outlined above is that it allows you to optimize your sensor assembly. Possible optimization objectives include size, performance, and cost. By quickly allowing you to see the effect of using a particular Hall IC with a particular magnet, it is possible to look at the performance limits of many alternatives, with minimal physical experimentation. This information can be used for selection of the most appropriate combinations of magnetics and Hall-effect IC for the application.

None of the above, of course, is a substitute for actually building a usefully large sample of devices, measuring their performance, and ultimately testing them in the end-application before committing the design to production. The general approach outlined above can merely provide a little bit of understanding into how a given design works. This in turn makes it easier to get to a point where you have a design that meets your cost and performance goals.

Chapter 7

Current-Sensing Techniques

While Hall-effect sensors are most frequently used to detect the proximity, position, or speed of mechanical targets, they are also useful for sensing electrical current. This is done indirectly, by measuring the magnetic field associated with current flow. Before getting into magnetic current sensing, let's talk about the more commonly used technique of resistive current sensing.

7.1 Resistive Current Sensing

Resistive current sensing exploits the voltage drop associated with an electrical current flowing through a resistor, as shown in Figure 7-1.

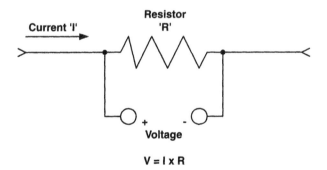

Figure 7-1: Current sensing with a resistor.

The voltage measured across a resistive current sensor is proportional to the current passing through the resistor, and is given by $V = IR$. Because resistors are inexpensive and are readily available in a number of forms, power ratings, and precisions (accuracies of 0.1% and better are readily obtainable), resistive current sensing is both popular and useful in many applications. Resistive current sensing does, however, have a few disadvantages. Among the more significant are:

- Lack of electrical isolation
- Common-mode voltage
- Voltage drop and energy loss

Lack of electrical isolation between the power-handling circuit and measurement circuit can be a major drawback with resistive current-sensing techniques. Because it is necessary to measure the voltage at the terminals of the resistor, the current-sensing instrument must become part of the circuit being monitored. For some applications, particularly ones operating at low voltage and power levels, this may not be an issue. For other applications, however, safety concerns will make it mandatory that the measurement circuit be separated from the power-handling circuit. This is often the case when monitoring current in systems that operate at line (117/220 VAC) voltage or higher. To provide isolation for a resistive current sensor requires the addition of significant amounts of electronics.

The requirement of measuring the voltage across the resistor can also become a problem, especially if the voltages at both of those terminals are significantly above the measurement system's ground reference. As an example, consider a 0.1Ω current sensing resistor in a 50V line. Each ampere of current will only result in 0.1V of signal across the resistor. For the case of 1A of current, the voltages at each side of the resistor with respect to system ground will be 50V and 49.9V. Discriminating this small differential signal (0.1V) riding on a large common mode signal ($\approx$ 50V) can be a difficult measurement problem. One way to deal with this situation is to add electrical isolation circuitry to the measurement system, and allow part of the measurement system to "float" at the higher voltage. This is essentially what happens when one uses a handheld DVM (digital voltmeter) to measure current; the meter electronics "float" up to whatever voltage is present at the input terminals.

Voltage drop and its associated energy loss is the third major problem with resistive current sensing. In order to sense current, one needs to develop a voltage drop across the resistor. This voltage drop is given by $I \times R$. What constitutes an acceptable voltage drop is determined by the application; usually the smaller, the better. For purposes of measurement accuracy, however, one typically wants as much voltage drop as possible. The designer must therefore make the appropriate trade-off between signal magnitude and the degree to which the measurement interferes with the system being measured.

The voltage drop across the measuring resistor is also associated with power dissipation, in the form of heat. Power dissipation is given by $P = I^2R$, and is significant for two reasons. The first is that this is energy that is lost from the system under measurement, and its loss results in reduced efficiency. The second reason is that enough power dissipated in a resistor will make that resistor hot, and may require special measures to keep it cool. For sensing small amounts of current (100 mA) resistor power dissipation may not be a major concern. For larger currents, such as those that might be found in a battery charger or large switch-mode power supply (>100A), power dissipation in a sense resistor can literally become a burning issue.

Now that we have discussed a few of the more important shortcomings of resistive current sensing, we can talk about magnetic current sensing. The main advantage of magnetic current sensing is that it doesn't interfere with the circuit in which the current is being sensed (at least for DC currents). Because one is measuring the magnetic field around a conductor, there is no electrical connection to that conductor. This automatically provides electrical isolation from the circuit being monitored. This also means that common-mode measurement effects disappear, because a magnetic current sensor doesn't care what voltage potential the line being measured is at, provided, of course, that it is within the designed safety limits of the sensor assembly (e.g., the insulation doesn't break down). Because a resistive component is not added in series with the circuit, there are no additional *IR* voltage losses or *I²R* power dissipation effects. This allows magnetic current sensors to be used to measure high current levels without excessive power dissipation.

7.2 Free-Space Current Sensing

Conceptually, a current sensor can be made by placing a linear Hall-effect sensor in close proximity to a current-carrying conductor. The sensor should be oriented so that the magnetic flux lines, which circle the conductor as shown in Figure 7-2, can be detected.

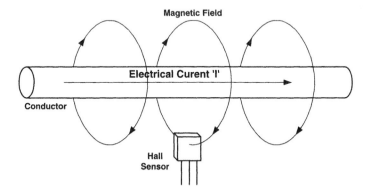

Figure 7-2: Current sensing in free space around conductor.

Assuming that nothing other than empty space surrounds the conductor, the conductor has a circular cross-section, and extends off in a straight line to infinity in both directions, the magnitude of the sensed field is given by Equation 7-1.

$$B = \frac{\mu_0 I}{2\pi r}$$ **(Equation 7-1)**

where r is the distance from the centerline of the conductor to the sensor.

While this scheme is simple and elegant, several difficulties can arise when try-ing to implement it in the real world. First, you don't get very much field for a given amount of current, at least for macroscopic conductor-to-sensor spacings. A 10-A cur-rent develops only about 2 gauss at a 1-cm distance. One consequence of this is that the sensor will be influenced by external fields. Considering that the earth's magnetic field is about ½ gauss, a current sensor implemented in this manner (with a 1-cm spacing) could experience as much as ±2.5A of error based just on the direction in which it is pointing (north-south-east-west). While potentially useful for current sensors intended for measuring large currents (hundreds or thousands of amperes), this approach may not be suitable for measuring smaller currents where extraneous fields could cause significant measurement errors.

A second difficulty with this scheme is that it is highly sensitive to positioning er-rors between the conductor and sensor. This becomes especially true when the separa-tion is small. To make an effective current sensor, it is necessary to tightly control the location of the sensor.

A third difficulty is that the sensitivity given above assumes the existence of an "infinitely long, straight conductor" (as the physicists like to say). Unfortunately, you can't buy this kind of conductor anywhere, and you will probably have to settle for one of the short, bent variety found in the real world. Conductor geometry will have some effect on the sensitivity of one's current sensor. This is bad news if the conductor can change shape or flex after installation and calibration.

One situation where it may be possible to overcome many of these problems is when both the conductor and the Hall-effect sensor are rigidly mounted on a printed-circuit board, as shown in the examples illustrated in Figure 7-3. By using the PCB trac-es to carry current, it is possible to tightly control the conductor geometry and spacing to the sensor. Note that the Hall-effect sensor is not placed over the conductor, but next to it. This is because the magnetic flux immediately over the conductor runs parallel to the surface of the board, and is therefore not coincident with the sensor's sensitive axis. The flux does, however, run perpendicular to the board past the edge of the conductor, and this is where the sensor should be placed, as shown in Figure 7-3a.

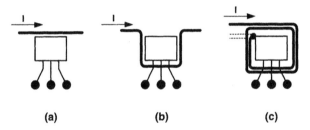

Figure 7-3: Current sensors on printed circuit boards.

When making a PCB-based current sensor, one can employ a number of techniques to get more field per unit current and increase the system's sensitivity. The first is to use as narrow a PCB trace as is possible consistent with safely handling the expected current flow. Flux lines resulting from a given current have a shorter path when looping around a narrow PCB trace than a thicker one, resulting in more field per unit of current. Note, however, that as you reduce the width of the PCB trace, its resistance increases and its maximum allowable current ratings will decrease. Additionally, if you encapsulate the PCB, the maximum current that can be safely handled may vary considerably depending on the thermal conductivity of the encapsulation material.

Another flux-intensification technique is to form a cul-de-sac into which the sensor is inserted (Figure 7-3b). This effectively superimposes the fields from the three sides of the cul-de-sac onto the sensor. One can also extend this idea further, and create loops (Figure 7-3c) in the PCB trace, and place the sensor in the middle. This geometry, will requires either a double-layered board or the use of jumpers. Be very cautious when using ordinary PCB plated-though "vias" for carrying high currents, as they can have relatively high resistance and become prone to failure when carrying high currents.

Note that, because the conductor geometries of the above examples have departed far from the physicists' ideal, the magnitude of the field is no longer given by Equation 7-1. Because of these more complex geometries, finite-element analysis (on a computer) or sophisticated mathematical analysis (by hand) are necessary to accurately predict the resulting magnetic fields.

Another issue, previously alluded to, that arises from using PCB traces for current sensing is that they have nonzero resistance. While this results in heating effects and imposes thermal limits on how much current can be safely handled, excessive resistance also can interfere with the measurement process. If you use a long-enough PCB trace, you may add enough resistance to the current-carrying circuit to interfere with its operation, and consequently reduce the accuracy of the current measurement. Adding series resistance into the measurement circuit reduces one of the primary benefits of using magnetic current sensors.

As a final note, electrical isolation can become a serious issue with PCB-based current sensors. Because the current-carrying traces are in close proximity to the sensor (and associated measurement electronics), sufficiently high voltage differences can cause current to flow between them. Typical leakage paths are through the PCB substrate (insulation breakdown), over the PCB surface (creepage), or through the air above the PCB (arcing). If safety or electrical isolation are a concern when implementing this kind of sensor, consult with the relevant safety agencies (e.g., UL, CSA, TUV) to obtain guidelines and standards for construction techniques suitable for various environments and applications.

7.3 Free-Space Current Sensors II

While using a Hall-effect sensor placed near a conductor as a current sensor may not be the most practical solution to most current-sensing problems, it can be useful in applications where very high current levels need to be sensed, and conductor geometry can be controlled. Increasing the effectiveness of potential designs requires that we take a look at some of the underlying physics in more detail.

Equation 7-1, describing the field around a conductor, is a special case of one of Maxwell's equations. A more basic relationship, valid for all geometries of current flow and magnetic field (in empty space), is given by Equation 7-2.

$$\oint B \cdot ds = \mu_0 I \qquad \text{(Equation 7-2)}$$

To illustrate what this means from a physical standpoint, consider Figure 7-4a, where a current-carrying conductor of arbitrary shape and size is completely enclosed by an arbitrary path around it. The only constraint is that the path must completely enclose all of the current. If one makes a single circuit of this path, integrating the magnetic field tangent to the path direction as one moves along, the integral of the field will be $\mu_0 I$.

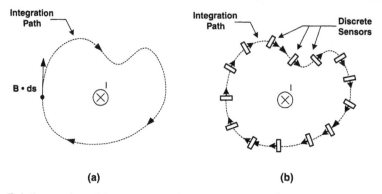

$$\text{(a)} \qquad\qquad\qquad\qquad \text{(b)}$$

Figure 7-4: Integration of flux around a closed path (a) and discrete approximation (b).

This integral may also be approximated by taking a large number of discrete measurements at evenly spaced points along the path (Figure 7-4b). Again, the axis of each measurement has to be tangential to the path at the point at which it is taken. The sum of all these measurements is proportional to the current enclosed by the path, regardless of the geometry of the current or the geometry of the path. Additionally, this aggregate measurement will be relatively insensitive to current sources outside of the sensor path, provided enough measurement points are used.

The relationship defined by Equation 7-2 is exploited by an inductive current sensor called a Rogowski coil (Figure 7-5), where a coil of many turns of wire is wound on a long, flexible carrier. To measure current, the carrier is wrapped once around the current-carrying conductor, and a voltage read from the coil. Because it is inductive, the Rogowski coil is only useful for measuring AC currents. Because of its physical flexibility, however, it is very useful for making measurements in places where physical accessibility may be limited.

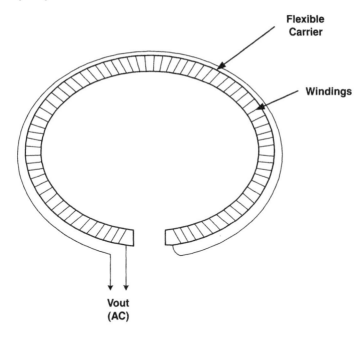

Figure 7-5: Rogowski coil AC current sensor.

While one could build a Hall-effect version of a Rogowski coil, using a large number of discrete Hall-effect sensors instead of wire windings, it would be an expensive proposition; good linear Hall-effect sensors cost considerably more than copper wire. Some of the benefit of the ring-of-sensors concept can also be obtained by using just a few devices. Consider the example shown in Figure 7-6, where four sensors are placed along a circle of radius r at 90° increments. Note that the orientation and polarity of each sensor is arranged to provide a positive response to magnetic flux flowing in a clockwise direction.

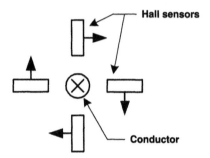

Figure 7-6: Free-space current sensor using four Hall-effect devices.

Although four sensors will not provide a very close approximation to the contour integral of Equation 7-2, they do make for a current sensor that is relatively insensitive to conductor position, at least when compared to the case where one uses a single sensor. By assuming that the central conductor is circular, infinitely long and straight, an analytic expression (Equation 7-3) can be derived for the total flux sensed by the four sensors as a function of conductor position.

$$B_T = \frac{\mu_0 I}{2\pi}\left(\frac{x+r}{\left((x+r)^2+y^2\right)} - \frac{x-r}{\left((x-r)^2+y^2\right)} + \frac{y+r}{\left(x^2+(y+r)^2\right)} - \frac{y-r}{\left(x^2+(y-r)^2\right)}\right)$$

(Equation 7-3)

Each of the terms represents the tangential field seen by each of the four sensors. Although Equation 7-3 is a closed-form solution, it doesn't provide much intuition into the system's behavior. Figure 7-7, a plot of the value of B_T as a function of x and y, yields a bit more insight.

When the conductor is located near the center of the four sensors, the response is relatively uniform; it is within ±2% out to a radius of $0.4r$. When one moves the conductor near any one of the four sensors, however, the response increases dramatically. When one leaves the circle, it drops equally dramatically towards zero. Additionally, consider the effect of a uniform, externally imposed field on the system. Extraneous uniform fields detected by any one sensor will be canceled out by the measurement from the sensor on the other side of the ring. Even for the case of nonuniform external fields (such as what might be obtained from a nearby magnet), the opposing sensors will still often provide some cancellation.

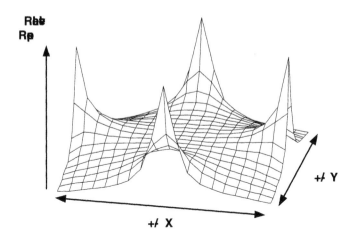

Figure 7-7: Response of four-probe current sensor as function of conductor position.

To illustrate this type of sensor, consider four A1301 linear Hall-effect sensors placed on a 2.5-cm diameter ring, mounted into a PCB to fix their positions (Figure 7-8a). A 1-cm hole is provided to pass a conductor through. The outputs are tied together through 10 kΩ resistors so that they are averaged together (Figure 8-8b). A 0.1-μF capacitor is also included for power supply decoupling.

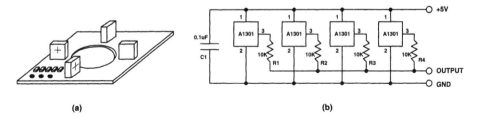

Figure 7-8: Example of ring sensor. Physical layout (a), schematic (b).

If one assumes the conductor is centered in the ring, the field-per-unit-current seen by all four sensors will be equal, and is given by:

$$\frac{B}{I} = \frac{\mu_0}{2\pi r} = \frac{4\pi \times 10^{-7}\,(\text{H/m})}{2\pi \cdot 0.0125\text{m}} = 1.6 \times 10^{-5}\,\text{H/m}^2\,(\text{T/A}) = 0.16\,\text{G/A}$$

(Equation 7-4)

Tying the outputs together through resistors has the effect of averaging them, so the overall sensitivity of this sensor is the same as the average of the sensitivities of the individual A1301's, nominally 2.5 mV/G. Multiplying this by the magnetic gain of the circuit gives an overall sensitivity of:

$$\frac{V_{OUT}}{I} = \frac{0.16G}{A} \times \frac{2.5mV}{G} = 0.4mV/A \qquad \textbf{(Equation 7-5)}$$

Because the A1301 will provide linear output for fields over the approximate range of ±800 gauss, this current sensor assembly will be able to measure current over a span of ±5000A (800G / (0.16G/A)). Because we only used four sensors to approximate a true integration, there will be significant sensitivity to current outside the ring (as indicated above), and the shape and path of the conductor as it passes through the ring will also influence the sensitivity as measured in volts/ampere.

7.4 Toroidal Current Sensors

Because free-space current sensors suffer from the dual problems of lack of sensitivity and susceptibility to outside interference, they are often not the best choice for most applications. Fortunately, there is a simple way to make highly sensitive current sensors that are very immune to external interference. This is accomplished by placing a high-permeability flux path around the conductor to concentrate flux at the sensor. Although many structures can be used to perform this function, toroidal flux concentrators are commonly used for this purpose, and for the purposes of this discussion are also amenable to simple analysis.

When one places a high-permeability toroid around a current-carrying conductor, the flux flowing around the toroid can be significantly greater than that in the space around it. When the conductor is centered in the toroid, the toroid's presence will not alter the shape of the field (circular, as shown in Figure 7-9a), but will cause significant increase in flux density (Figure 7-9b) when one enters the interior of the toroid.

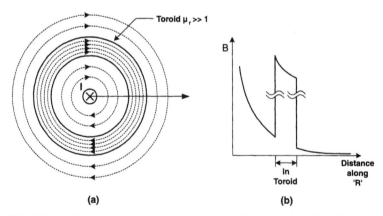

Figure 7-9: Magnetic field in and near toroid (a) and intensity profile (b).

For a toroid constructed of an idealized linear magnetic material, where μ_r is constant with respect to magnetic flux, the field in the toroid can be calculated by:

$$B = \frac{\mu_0 \mu_r I}{2\pi r} \qquad \textbf{(Equation 7-6)}$$

where r is the radius of the toroid at the point of interest and μ_r is the relative permeability of the toroid. A material's relative permeability can be thought of as how well it can "conduct" a magnetic field as compared to free space.

When the relative permeability of the toroid is very high (>100–1000), the field in the toroid becomes less dependent on exact placement of the conductor. This occurs for two reasons. First, magnetic flux density can't suddenly drop without diverging. This means that the flux flowing around the toroid is constrained to flow around the toroid; for the most part it can't take a shortcut through the air in the middle. A small amount of the flux does do this, however, but the more permeable the toroid is, the smaller the portion of leakage flux escapes (exaggerated in Figure 7-10). This containment effect causes the flux to be more or less uniform in the toroid, with a higher intensity near the inner diameter.

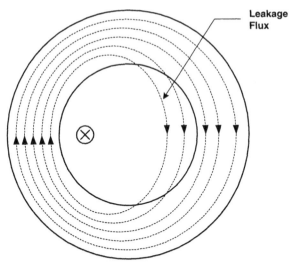

Figure 7-10: Off-center current causes leakage flux to escape toroid.

Because the magnetic flux is constrained to a path, the integral of the flux has to be proportional to the current surrounded by that path. This makes the flux essentially independent of the placement or geometry of the current, as long as it passes through the center of the toroid.

7.5 Analysis of Slotted Toroid

The previous analysis described how magnetic fields behave in a high-permeability closed toroid. To measure the magnetic field, however, requires that a sensor be inserted into the flux path. This can be done most easily by cutting the toroid so that it becomes a C-shaped structure. This slot must be cut all the way through the toroid (Figure 7-11a), flux will tend to seek the shortest path through the uncut section and avoid jumping across the gap (Figure 7-11b).

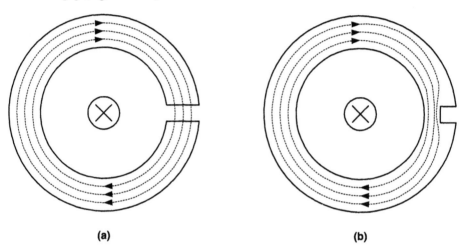

(a)   (b)

Figure 7-11: Preferred (a) and nonpreferred ways (b) to slot a toroid.

Because the flux must be continuous as it crosses the faces of the toroid forming the gap, it must be the same as the flux inside the toroid. In actuality, the flux will diverge (fringe) as it crosses the gap, so the intensity will decrease a bit. If the gap length is short compared to the linear dimensions of the gap faces, however, this decrease will be relatively small. Conversely, the field strength can drop significantly if the gap width is large compared to the gap face dimensions. When the gap width (g) is small compared to the toroid circumference ($g \ll 2\pi r$) , the flux can be approximated by:

$$B = \frac{\mu_0 \mu_r I}{2\pi r + g\mu r} \qquad\qquad \textbf{(Equation 7-7)}$$

Furthermore, if $g\mu_r \gg 2\pi r$, which is often the case for a high-permeability toroid, then the expression can be further simplified to:

$$B = \frac{\mu_0 I}{g} \qquad\qquad \textbf{(Equation 7-8)}$$

This result means that, if the permeability of the toroid material is sufficiently high, the amount of flux in the toroid is controlled by the gap width g.

Another way of describing a toroid's efficacy at converting current into magnetic field is by its magnetic gain, (A_M) expressed in gauss/ampere. From Equation 7-8, the gain of a toroid of gap g can be expressed:

$$A_M = \mu_0/g$$ (**Equation 7-9**)

Several manufacturers produce slotted toroids as standard items for current-sensing applications with 0.062" ($\approx$1.5 mm) gap widths; this width accommodates several types of commonly available Hall-effect IC packages. From Equation 7-9, one would expect a gain of 8.4 gauss per ampere. In practice, because of finite permeability, flux leakage, and fringing effects in the gap, one can usually expect to see actual magnetic gains ranging anywhere from 6–8 gauss per ampere.

7.6 Toroid Material Selection and Issues

Although we just finished showing that the gap width is the dominant factor in the sensitivity of a flux-concentrating toroid, there are many material issues that must be addressed in order to get a current sensor to behave as intended.

The magnetic characteristics of the "soft" magnetic materials used for current-sensor toroids can be described by a B-H curve, as are permanent magnet materials. The major difference is that for soft magnetic materials we are more interested in their quadrant I characteristics than those of quadrant II. The B-H curve for a "typical" ferrite core material is shown in Figure 7-12.

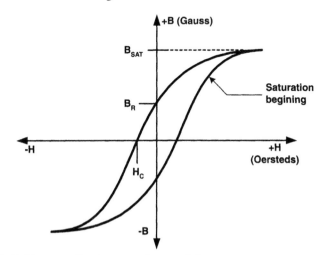

Figure 7-12: B-H curve of soft magnetic material.

Some of the key characteristics that can be derived from this curve are relative permeability (μ_r), remanent flux (B_r), coercivity (H_c), and saturation flux (B_{sat}). Relative permeability describes the relationship between how much coercive force (H) is applied to the material, and how much flux density (B) will be induced in the material. Note that when B approaches B_{sat}, the effective permeability will begin to fall, as increases in H will no longer produce comparable increases in B.

Once the material has been driven into saturation and the field removed, it will keep a certain remanent flux density B_r in the absence of any coercive force. Note that this remanent flux will only appear in a closed magnetic circuit composed of the material, such as a continuous toroid. Cutting a slot in the toroid will interrupt this path and greatly reduce the flux density. This is why a slotted toroid with a B_r specified as a few hundred gauss will only exhibit a field of a fraction of a gauss in the slot.

After the material has been saturated in the positive direction, a small amount of negative coercive force H is necessary to drive the flux B back to zero. This amount of coercive force is referred to as H_C, the material's coercivity, and is characterized in oersteds (Oe). For a soft ferrite material that might be suitable for a current sensor toroid, where one wants the flux to easily return to zero, very small coercivities (< 1 Oe) are desirable. This is in contrast to the very high coercivities (>10,000 Oe) found in many permanent magnet materials where one wants the material to stay permanently magnetized despite any external fields.

For most current-sensing applications, one will want to select a material with a saturation flux density significantly greater than the flux densities it will see in operation. This is because saturation begins to occur gradually in many materials, with permeability starting to decrease as flux levels increase past a certain point. This effect will manifest itself as gain and nonlinearity errors in the final sensor assembly. In some cases, toroid manufacturers will make your life simpler in this department by specifying the maximum number of ampere-turns that can be applied to a given toroid before it begins to saturate. One problem with relying on a maximum ampere-turns measure is that the provided figure may represent operation sufficiently far up the saturation curve to cause unacceptable linearity errors in your application.

Permeability also can vary significantly over temperature. The main concern here is that the permeability does not drop to the degree that it begins to affect the amount of flux developed in the toroid. This issue can be addressed by using materials that exhibit at least some minimum permeability over the operating conditions under which you will be operating your current sensor.

7.7 Increasing Sensitivity with Multiple Turns

There are three ways of getting more sensitivity out of a toroidal current sensor. The first is to use a more sensitive magnetic transducer. The second is to use a narrower airgap (and a thinner transducer to fit). Finally, the simplest method of all is to loop the conductor through the toroid a multiple number of times. Looping a conductor carrying *I* amperes through the toroid *N* times causes *NI* total current to pass through the toroid.

One frequent question is that of how one counts turns. The answer, to a first-order approximation, is quite simple. A turn is the passage of the conductor through the center of the toroid. It largely doesn't matter what you do with the wire elsewhere. For this reason, turns only occur as integers; you can't have half of a turn. Because of the nonideal characteristics of any given current sensor's magnetic circuit, this is not completely true; conductor placement can and does have a small effect on a current sensor's sensitivity. In many if not most situations, however, one will find that counting turns as wire passages through the toroid will give acceptable results.

7.8 An Example Current Sensor

Because current sensing is a popular application for linear Hall-effect sensors, many vendors supply suitable off-the-shelf components for implementing them. This makes the development of simple current sensors relatively straightforward.

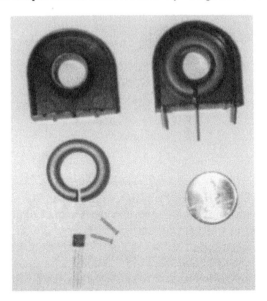

Figure 7-13: Photograph of unassembled and assembled prototype current sensor.

Figure 7-13 shows both unassembled and assembled views of a prototype current sensor. To make a finished unit, all that is needed is to "pot" the toroid and sensor with a suitable encapsulation material. The toroid is a Philips model TX22/14/6.4-3C81, about 7/8" in diameter, made from a material (3C81) with a permeability of about 3000, and good linearity up to around 800 gauss. A 0.062" gap was cut in the toroid, so it will provide about 6–8 gauss/ampere-turn, and will offer linear operation up to around 100–120 ampere-turns. The Hall-effect sensor is an Allegro Microsystems A1373, which provides a linear output with a sensitivity that can be programmed over from 0 to 7 mV/G. Even the enclosure is off-the-shelf, manufactured by Robison Electronics specifically for current-sensor applications. If the Hall-effect sensor is programmed to a sensitivity of 5 mV/G this current sensor will have a gain of about 30 mV/A, with saturation points of roughly ±80A, limited by the Hall-effect sensor's output saturation characteristics. By choosing a programmable sensor such as the A1373, this assembly can be adjusted at manufacture to a high degree of uniformity in both overall gain and offset error.

7.9   A Digital Current Sensor

Although we have only discussed linear-output current sensors to this point, it is also possible to make current sensors that provide an ON/OFF logic output if a current threshold is exceeded. By putting a switched digital Hall-effect sensor in the toroid gap, one can make a threshold-sensitive current sensor. This sensor turns on when the current exceeds a given threshold (I_{OP}), and turns off when the current drops below a second, lower threshold (I_{RP}). When constructed with a Hall-effect switch with operate and release points B_{OP} and B_{RP}, the resulting current sensor will have I_{OP} and I_{RP} points given by:

$$I_{OP} = B_{OP}/A_M \qquad\qquad\qquad\qquad \textbf{(Equation 7-10)}$$

$$I_{RP} = B_{RP}/A_M \qquad\qquad\qquad\qquad \textbf{(Equation 7-11)}$$

where A_M is the magnetic gain (gauss/ampere) of the toroid.

For example, if a digital Hall-effect switch with a B_{OP} of 200 gauss, and a B_{RP} of 150 gauss were used with a toroid with a gain of 6 gauss/ampere, the resulting assembly would switch on (operate) when the current exceeded (200/6) = 33.3A and switch off (release) when the current dropped below (150/6) = 25A. Note that, because the polarity of the magnetic flux in the toroid will be proportional to the polarity of the current, this device will only turn ON in response to current of a given polarity.

Because most Hall-effect switches have a fairly wide range of B_{OP}/B_{RP} points from unit-to-unit, it would be difficult to make devices with tightly controlled current thresholds. For this application, a Hall-effect switch with programmable B_{OP}/B_{RP} points, such as Allegro Microsystem's A3250, would allow you to trim each sensor on a unit-by-unit basis in production to provide uniform trip-point currents.

7.10   Integrated Current Sensors

As we showed in Section 7.5, the primary factor controlling sensitivity of a toroidal Hall-effect current sensor is the width of the gap into which the sensor is inserted. Most commercially available integrated devices tend to have package thickness ranging from 0.050" to 0.060". This limits the maximum amount of field one can get out of the toroid to roughly 8–10 gauss per ampere-turn.

The actual silicon die on which the sensor and electronics is fabricated is in the order of 0.010" thick. If it were possible to dispense with the device's packaging, it would be possible to realize much more sensitive current sensors. Handling bare dice, however can be difficult, as they tend to be fragile, sensitive to contamination, and require specialized equipment to attach interconnections. For these reasons it will not be practical for most manufacturers to build current sensors using bare dice.

Another alternative is to integrate the current sensor magnetic structures into the sensor packaging. Allegro Microsystems took this approach with their ACS750 series of integrated current sensors [Stauth04]. These devices contain all of the key components of a toroidal current sensor: a flux concentrator, a linear sensor, and also integrate the current-carrying conductor. Figures 7-14a and 7-14b show a photograph and cross-sectional schematic view of this device.

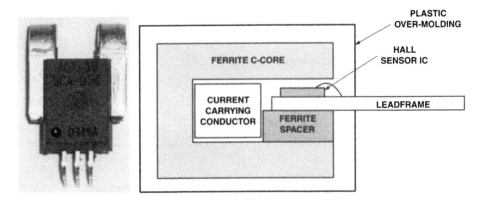

Figure 7-14: ACS750 photograph (a) and cross-sectional view (b).

In the AC750, the Hall-effect sensor die is mounted to the leadframe as it is in more conventionally packaged integrated sensors. The leadframe is then positioned in the gap of a ferrite c-core, along with a ferrite spacer to reduce the effective gap length. A heavy, low-resistance conductor for carrying the current to be measured is also included. Finally, all of these components are overmolded with an epoxy compound that both locks everything in place and protects the integrated circuit and wire bonds. The resulting assembly provides a high-sensitivity current sensor capable of measuring currents up to 100 A that can be inserted into a printed circuit board.

Allegro Microsystem's ACS704 is a recently introduced and related integrated current sensor that is implemented in a SOIC-8 IC package. Again, as in the case of the ACS754, the device provides a conductor for the current to be measured and includes an integrated Hall-effect sensor for measuring the resulting magnetic field. The primary advantage provided by this device is its small physical size, which is approximately 0.3" × 0.3" × 0.06". Because of limitations of the SOIC-8 package, the maximum currents that can be measured with this device are in the range of 5–14A.

7.11   Closed-Loop Current Sensors

Although it is useful for many applications, the toroidal current sensor presented has several limitations that restrict its use as a precision measuring device. Specifically, linearity and gain errors resulting from both the toroid and Hall-effect sensor severely limit the accuracy attainable with this type of device.

In many electronic systems, linearity and gain errors are corrected by using negative feedback, and this technique can also be employed in magnetic current sensors. Figure 7-15 shows how this is done.

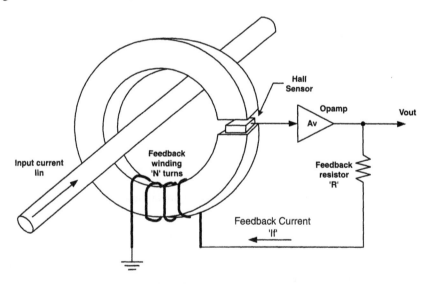

Figure 7-15: Closed-loop current sensor.

Qualitatively, this sensor operates by attempting to balance the input current with an opposing feedback current. When the net current passing through the toroid is zero, the flux through the toroid is also zero. The feedback current is coupled into the toroid through a separate conductor arranged to develop a magnetic field opposed to that of the current being measured. At first glance this would seem to be a serious drawback, as a significant amount of current could be required to perform this cancellation. However, because it is possible to effectively "multiply" a given current through the use of multiple turns, a small current through a large number of turns can be used to balance out a large current through a single turn. This means that, with enough turns in the feedback winding, only a modest amount of feedback current is required. As an example, if the feedback winding has 1000 turns, only 10 mA of feedback current will be required to balance out 10A of input current, which is only present on a single turn.

One significant advantage of using a current-balancing approach is that, since the net flux in the toroid is zero, the range over which input current can be measured is limited only by the available feedback current (multiplied by turns). It is therefore possible to make current sensors that can sense extremely large currents without magnetically saturating the toroid or the Hall-effect sensor.

The next part of the current sensor is the Hall-effect transducer. The only exacting requirement for this device is that it be able to detect when the magnetic flux is zero. Maintaining an exact transducer gain is not critical. Offset errors, on the other hand, are still important, as they prevent one from knowing when the toroid flux is truly zero.

At this point, one could make a manual current-measuring device using a readout from the Hall-effect sensor to indicate zero, and a current source controlled manually with a knob to provide feedback current. One would adjust the current source until the sensor indicated zero flux, read the amount of current required, and multiply by the number of turns in the feedback coil. While this could be done for a laboratory measurement, it is still inconvenient and totally unnecessary, as this function can be performed by a high-gain amplifier placed in a feedback loop.

In the feedback loop employed by this current sensor, the difference between the current to be measured and the feedback current results in an error signal. The amplifier in this circuit attempts to reduce this error signal to zero, by altering its output appropriately. When the error signal is zero (or very close), the output of the amplifier is proportional to the current being measured.

Systems constructed around feedback loops are elegant because they allow minimization of the nonideal effects of many of the components. The DC steady-state behavior of a loop such as this one is straightforward to analyze. The first step is to represent the whole system in block diagram form, as shown in Figure 7-16.

The first operation is the summation of the input current with the feedback current. Next, this is multiplied by the toroid's magnetic gain A_M (in G/A). The next stage is the transducer gain A_H (in V/G) followed by the op-amp voltage gain A_V (in V/V). The resulting output signal is then converted to a current through the feedback resistance $(1/R)$, and finally multiplied by the number of feedback windings (N) before completing the loop. A closed-form expression (Equation 7-12) may be derived that relates V_{OUT} to I_{IN}.

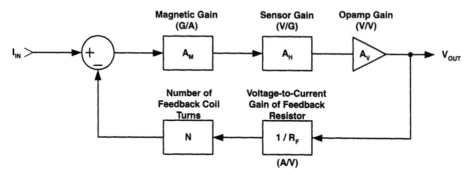

Figure 7-16: System diagram of feedback current sensor.

$$\frac{V_{out}}{I_{in}} = \frac{A_M A_H A_V}{1 + \dfrac{A_M A_H A_V N}{R}} \qquad \text{(Equation 7-12)}$$

In this equation, all of the factors discussed above contribute to the overall gain of the closed-loop current sensor. One of the near-magical properties of negative feedback, however, is that if one makes the gain of the op-amp sufficiently large ($A_V \gg A_M A_H N/R$), the relationship between V_{OUT} and I_{IN} can be approximated as:

$$\frac{V_{OUT}}{I_{IN}} = \frac{R}{N} \qquad \text{(Equation 7-13)}$$

The only terms remaining are the value of the feedback resistor and the number of feedback turns. Both of these can be controlled to a high degree of accuracy.

The benefits of feedback, however, do not come without a price; neither does feedback magically solve all problems. As mentioned before, the coercivity of the toroid and the offset of the Hall-effect sensor still will contribute offset errors to the same degree as they would in an open-loop current sensor. A more complex issue, however, is that of dynamic stability. By adding a high-gain feedback loop to the system, we have introduced the possibility that it will exhibit oscillatory behavior. Small amounts of instability can manifest themselves as "overshoot" or "ringing" in response to a quick change in input current—with adverse effects on dynamic measurement accuracy. Major instabilities can result in uncontrolled oscillation, where the current sensor outputs a large sinusoidal waveform, even when there is no input. The dynamic behavior of the system will be a complex function of the dynamics of all the components included in the feedback loop. A detailed analysis of system stability issues, however, is beyond the scope of this book.

Chapter 8

Speed and Timing Sensors

The ability to measure the speed or position of a rotating shaft is necessary to the proper functioning of many types of machinery. Speed sensors, timing sensors, and encoders are used in applications as varied as automobile ignition controls, exercise equipment, and CNC machine tools. An example of a typical Hall-effect speed sensor is shown in Figure 8-1. When a toothed steel target (such as a gear) is rotated past its flat end, this device provides a train of output pulses, one for each passing target feature.

Figure 8-1: Geartooth sensor assembly (courtesy of Cherry Electrical Products).

8.1 Competitive Technologies

Speed sensing can also be performed by sensors based on technologies other than the Hall effect. Among the more commonly used alternative technologies are optical, variable reluctance (VR), and inductive proximity.

Optical sensors can be used to detect speed by having features on a rotating target either interrupt or reflect a beam of light passing from an emitter (LED or laser) to a detector (phototransistor). Optical sensors exist in a wide spectrum of forms and prices, ranging from $0.50 opto-interruptor assemblies to high-resolution optical encoders costing hundreds or even thousands of dollars.

Variable reluctance sensors operate magnetically, and consist of a coil of wire wound around a magnet. As ferrous target features pass by the face of the sensor, they induce flux changes within the magnet, which are then converted into a voltage in the coil. VR sensors have the advantage of being very inexpensive and rugged. One application is in automotive antilock braking systems (ABS).

Inductive proximity sensors, also referred to as ECKO (eddy-current killed oscillator) sensors are frequently used as industrial automation components. These devices work by sustaining an oscillation in a high-Q LC circuit formed from a capacitor and a sensing inductor. The magnetic flux from the sensing inductor is allowed to pass to the outside of the sensor, through a detecting surface. When a conductive target is brought near the detecting surface, it absorbs energy from the magnetic field and damps the oscillation. Subsequent circuitry then reports target presence or absence based on the status of the oscillation.

While each of these alternative technologies has characteristics that suit it to particular applications, each also has its drawbacks. No single type of speed sensor, including Hall-effect devices, is the best choice for every application. Table 8-1 summarizes a few of the advantages and drawbacks of each of these technologies.

Table 8-1: Speed sensing technology comparison.

Technology	Advantages	Drawbacks
Hall effect	Hot, dirty environments No minimum speed (sometimes) Digital output	Requires ferrous targets
Optical	Fine spatial resolution Inexpensive Very fast (>100 kHz) Digital output	Susceptible to contamination Limited temp. range
VR	Inexpensive Hot, dirty environments Fast (>10 kHz)	Requires ferrous targets Minimum sensing speed Additional signal processing needed
Inductive (ECKO)	Hot, dirty environments Non-ferrous metal targets Digital outputs	Slow (<1 kHz) Low spatial resolution

8.2 Magnetic Targets

The most elementary Hall-effect speed sensor is one based on sensing the passing of a magnetized target affixed to a shaft. The magnetized target may take the form either of a number of discrete magnets, with pole faces presented to the sensor, or as a number of poles magnetized onto a single ring magnet. The decision whether to use a number of discrete magnets or a ring magnet will, as usual, be driven by functional, environmental, mechanical, and economic considerations.

If one chooses to use discrete magnets, there are two fundamental configurations available: one where each magnet presents the same pole to the sensor, and one where successive magnets present alternating poles. Figure 8-2 illustrates these two operating modes.

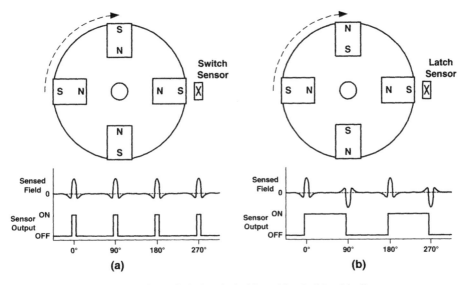

Figure 8-2: Use of Hall-effect digital switch (a) and latch (b) with discrete magnets.

When all the magnets are aligned so as to present the same pole to the sensor (Figure 8-2a), one will want to use a Hall-effect digital switch as a sensing element, with an operate point (B_{OP}) low enough to sense the field available at the operating airgap. Because this configuration is essentially identical to the slide-by mode presented in Chapter 6, the flux seen by the sensor will make a slight negative excursion as the magnet passes, assuring that a switch will be turned off. The resulting output will be a series of short pulses, each pulse corresponding to the passing of a magnet. The duty cycle of the output waveform is dependent on the following factors:

- Spatial size and separation of magnets.
- Magnet material and geometry

- • Magnet-to-sensor operating airgap
- • Sensor operate and release points

Because latch-type digital Hall-effect sensors can usually be obtained with lower maximum operate points than switch-type devices, it is possible to use a latch-type device to increase operating airgap. A latch, however, requires a negative field to turn it off. Mounting the magnets so that successive magnets present alternating pole faces (Figure 8-2b) meets this requirement. In exchange for higher operating airgap, however, one obtains a single output pulse for every two magnets that pass the sensor. This is because one magnet will turn the latch on, while the other magnet will turn the latch off. One additional advantage from this operating scheme is that, since one is controlling the turn-on and turn-off points by magnet position, it is possible to control the duty cycle of the output (assuming a constant shaft speed). Moreover, for the case of well-matched magnets evenly spaced around the target and symmetric latches (B_{OP} = B_{RP}), the duty cycle will remain nearly constant at 50% over a wide range of operating airgaps. One caveat with this approach is that if the latch's switchpoints are sufficiently low or the airgap is sufficiently small, the magnet's return field may inadvertently cause the latch to switch just after a magnet passes, instead of waiting for the next magnet.

While inserting a number of discrete magnets into a target can be economical for low-volume and specialty applications, ring magnets are often a better choice when manufacturing a sensor system in high volume. A ring magnet is a homogenous piece of permanent magnet material into which a number of poles have been magnetized (Figure 8-3a).

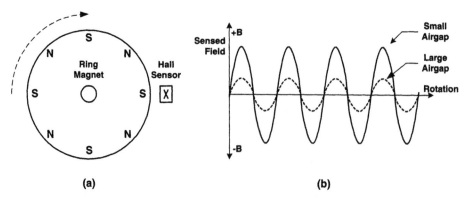

Figure 8-3: Ring magnet (a) and flux output vs. rotation (b).

One advantage of using a ring magnet is that it is possible to economically magnetize a high number of poles into one. Pole-counts ranging from 10–20 per linear inch of circumference are commonly achieved. For a 1-inch diameter magnet, this could mean 64 poles, or 32 pulses per revolution. For electronic speed-control systems high pulse-per-revolution counts are often useful for providing more accurate and consistent speed control.

Another characteristic of ring magnets, at least for ones where the poles are small relative to the operating airgap and are adjacent to each other (there are no unmagnetized regions in between the poles), the magnetic signal presented to the sensor tends to be somewhat sinusoidal as a function of rotary position. One consequence of this is that, if one uses a symmetric latch as a sensor, one will obtain a near-constant duty cycle (50%) over most of the operating airgap. A constant input duty cycle often makes the design of associated electronics simpler. There will also be a significant phase shift (½ pole width) as one moves the sensor away from the surface of the ring magnet out to maximum airgap. While this effect is inconsequential for many speed-sensing applications, it can pose a problem when one needs to reference an absolute point on the target.

One class of applications where magnet-sensing speed sensors are especially useful are those in which only a small amount of torque is available to move the target. Of all the Hall-effect-based speed-sensing methods to be presented in this chapter, magnet-based speed sensors do not put any mechanical torque load on the target, allowing it to spin freely. One case where this is essential is in turbine and paddlewheel-type flow meters, where the speed of a turbine or paddlewheel is proportional to the speed of fluid flowing through it. Discrete magnets can be embedded into the paddlewheel assembly (Figure 8-4) and sensed externally through the meter housing. In this application, the torque load and "cogging" caused by a geartooth sensor could provide unacceptable measurement errors, or even cause the paddle to stick in one position, particularly at low flow rates.

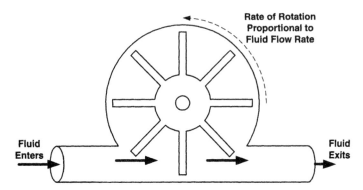

Figure 8-4: Paddlewheel type flow meter.

8.3 Vane Switches

The vane switch presented in Chapter 6 can also be used as a speed sensor. Because vane switches can provide good unit-to-unit repeatability in their mechanical actuation points, they have found widespread application as automotive ignition timing sensors. In this application, the vane sensor is used with a circular target presenting a number of steel flags that can pass through the vane switch's throat. Common target geometries include toothed wheels and notched cups, such as those shown in Figure 8-5. The actual size of targets such as those shown would be between 2 and 3 inches in diameter.

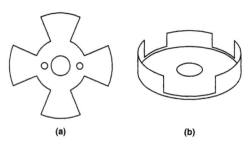

(a)   **(b)**

Figure 8-5: Disk (a) and cup (b) vane targets.

Although good unit-to-unit trip-point accuracy ($< \pm 1$-mm linear travel) is obtainable on the leading and trailing edges of the target, the number of vanes on a target can be limited. Because the dimensions of the vane flags and gaps between flags must be comparable to the dimensions of the vane-switch's throat, it is difficult to obtain high pulse counts from reasonably small vane interrupter targets.

What makes vane sensors attractive for speed-sensing applications in preference to ring-magnet schemes? The first advantage is that the target is a steel stamping; in high volumes these are extremely inexpensive to make. Secondly, a steel stamping is very rugged, resisting temperature, mechanical shock and many chemicals encountered inside automotive power train assemblies. Ring magnets tend to be made from either polymer-bonded or sintered materials. Bonded materials can have limited temperature range and solvent (gasoline, hot transmission fluid) compatibility issues. Sintered materials tend to be brittle and may not hold up under shock or repeated temperature-cycling conditions. Also, because a vane sensor only needs a small amount of magnetic material (which can be overmolded for environmental protection) the vane switch itself can be made at relatively low cost. The economy provided by the vane switch and its associated target is often the primary reason for using it in a given application.

Because vanes switches can exert a significant mechanical force on a passing target flag, they will tend to strongly detent or "cog" the shaft to which the target is attached. This can be an issue for two reasons. If the target is not driven with enough force to pull the target flags smoothly through the vane switch, then the target speed can vary as it rotates, or in the worst case it can become stuck. Another more subtle issue is that the force exerted by the vane switch on the target flags will tend to mechanically "pluck" them, resulting in acoustic noise. While this may not be a problem at low speeds, it can create annoying amounts of noise when the vane rate runs into mid-range audio frequencies (500–5000 Hz) or when the vane rate hits a resonant frequency of part of the system, in which case the resulting noise can be especially loud. While acoustic noise may not be an important consideration in some applications, it can be a major drawback in applications such as a household appliance or a piece of office equipment.

8.4 Geartooth Sensing

Often, it is not desirable to add a ring magnet or vane target to a system in order to sense its speed. In many cases this is because the speed sensor is designed in as the last part of the assembly, and an additional target cannot be accommodated without significant re-design and retooling. It then becomes desirable to be able to sense speed from features that either already exist or are easy to add to a rotating member, such as gears, pinions, keyways, and holes. If the available target features are ferromagnetic and have appro-priate geometries, Hall-effect speed sensors can often be used to detect them. Since ferromagnetic steels are ubiquitous in modern machinery, finding or making a suitable target is usually straightforward.

Detecting an unmagnetized steel target, however, presents a completely different set of challenges than those of detecting a magnet. When detecting an unmagnetized target, the sensor assembly must both provide a magnetic field and discriminate pertur-bations in that field caused by target features. A typical "geartooth sensor" will consist of a Hall-effect sensor IC placed on the face of a bias magnet, as shown in Figure 8-6a. The approach of a ferrous target (Figure 8-6b) intensifies the normal flux at the face of the magnet. Interpreting differences in flux patterns between the target present and target absent states is where the challenge of making a good geartooth sensor lies.

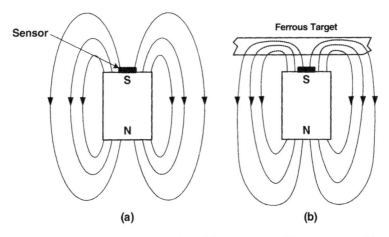

Figure 8-6: Flux around magnet (a) is altered by presence of ferrous target (b).

Because detecting presence or absence of a target can be difficult, many geartooth sensors are designed only to detect target features as they move past them. A sensor of this type may or may not register the presence of a target when the power is first turned on, after a power interruption, or when the target has not been moving for a short time. The ability to discriminate the presence of a stationary target feature is often called true proximity detection or power-on recognition. For many speed-sensing applications, lack of true proximity detection does not pose a significant problem. The lack of this feature, however, makes most types of geartooth sensors completely unsuitable for use as proximity detectors. When misused as a proximity detector, a geartooth sensor lacking true proximity detection capability can exhibit erratic and unpredictable behavior.

8.5   Geartooth Sensor Architecture

While there are literally dozens of schemes available for making Hall-effect based geartooth sensors, most of them fall into one of a few major categories. These categories are based on the features of the magnetic field being measured, and how the resulting measurements are interpreted into an output signal.

When measuring magnetic flux density in an effort to discriminate a target, one can either look for a magnitude or for a spatial gradient. While magnitude is readily measured with a single Hall-effect transducer element, spatial gradient cannot be measured directly; it must be approximated by subtracting the measurements from two independent, but closely spaced, transducer elements (Figure 8-7). Magnitude detection schemes are often called *single-point*, while gradient detection schemes are often called *differential*.

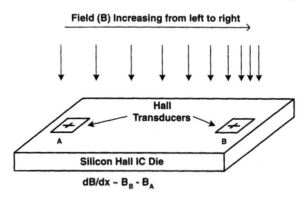

Figure 8-7: Two Hall-effect transducers on one die are used to measure flux gradient.

Both magnitude and gradient detect different characteristics of a given target. Changes in flux magnitude are useful for indicating the presence of the body of a target. One obtains the greatest magnitude response when the sensor is directly over the body of a target feature. Gradient detectors, on the other hand, respond to discontinuities in the target, such as the edges of gear teeth or the edges of holes. Each of these characteristics is useful in different applications.

The second dividing line is the method used to interpret the transducer signal. The basic idea in all schemes is to compare the transducer signal to a threshold and report presence or absence based on the results of the comparison. The distinction lies in whether the threshold is constant (a static threshold) or is allowed to vary over time (a dynamic threshold). While static thresholds offer conceptual simplicity, sensors based on the use of dynamic thresholds can offer significantly improved performance, ease of use, and applications flexibility, by allowing the sensor to adapt itself to the characteristics of the target being sensed.

The combination of magnitude (single-point) versus gradient (differential) sensing and fixed vs. dynamic thresholds yields four possible classes of Hall-effect-based speed sensors:

- Single point/fixed threshold
- Single point/dynamic threshold
- Differential/fixed threshold
- Differential/dynamic threshold

In the following sections, we will discuss the operation and examples of each of the above types of sensors.

8.6 Single-Point Sensing

In a single point geartooth sensor, only one Hall transducer is used to measure the magnitude of a magnetic field. To maximize the signal from the transducer, it is common to place the Hall-effect transducer between the target being detected and the pole-face of a magnet, as shown in Figure 8-8a. Typical magnetic responses, as seen by the sensor at various sensor-to-target airgaps, are shown in Figure 8-8b.

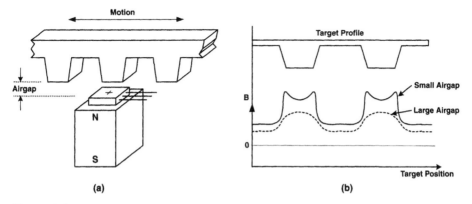

Figure 8-8: Pole face sensor (a) and response (b).

The magnetic response of this example shows several characteristics that are important to understand if intending to implement or use geartooth sensors using this magnetic configuration. The first is that the peak-to-peak signal is a function of airgap. The peak-to-peak signal allows one to distinguish target features from nonfeatures. In the case of a gear, the positive peak occurs when the sensor is over a tooth, and a negative peak occurs when the sensor is over the gear's root (valley between the teeth). The peak-to-peak signal drops rapidly, with a near inverse-exponential characteristic as a function of airgap. Because of this exponential fall-off rate, it is difficult to obtain greatly increased operating airgaps by using improved magnetic materials alone. A factor of three increase in net flux density (which might be provided by replacing an Alnico magnet with a Samarium-Cobalt magnet) will be unlikely to provide a 3× increase in operating airgap.

The second important signal feature is the baseline flux value measured when the sensor is positioned between teeth. This has two key characteristics, the first of which is its high value in relation to the peak-to-peak signal. The second characteristic of the baseline is that it tends to decrease as effective airgap increases, although by a much smaller amount than the peak-to-peak signal decreases. Targets with shallow gaps between teeth, and/or fine-pitched tooth spacing tend to have smaller differences between their baseline and peak flux densities than targets with larger tooth pitches and deeper gaps.

The high value of the baseline can pose a problem, especially for those sensors that must be able to provide true proximity detection. Remember that the sensor does not have the luxury of looking at the entire response curve shown in Figure 8-8b; it must make its target-present/absent decision based solely on its view of one or two closely spaced points on that curve. For the case of a small peak-to-peak signal variation riding on top of a much larger baseline reading, target determination can be difficult. Finally, a high magnetic baseline places addition requirements on the dynamic range of any signal-processing circuitry following the transducer. Subsequent circuitry must be able to accept high incoming signals without saturating, while still maintaining enough sensitivity to detect small changes in those signals.

A third characteristic of the magnetic response of the sensor is overshoot. This effect usually occurs at small effective airgaps (e.g., < 0.04"), and is caused by flux concentration at the corners of a target profile. This effect occurs because a corner or other sharp transition at the edge of a target feature can act as a better flux concentrator over a small region than the body of the target feature. Overshoot can be a problem because it often manifests itself by causing a geartooth sensor to output multiple pulses for each target feature.

Square-cut gear teeth, and other target features with sharp transitions, are especially prone to contribute to overshoot effects. Whether or not overshoot constitutes a problem depends on two factors. The first is whether the sensor is even sensitive to overshoot; some types, such as fixed-threshold single point, may not be if the trip points are adjusted to levels appropriate for the working airgap. The second factor is airgap.

Overshoot effects decrease rapidly with increased airgap, so even if the particular sensor doesn't work well in the presence of overshoot, specifying a minimum working airgap for a given target may solve the problem. Finally, adding chamfers or radii to target features can often eliminate overshoot in the magnetic signal.

8.7 Single-Point/Fixed-Threshold Schemes

One can make a useful geartooth sensor by simply comparing the signal resulting from the fixed-point configuration shown above against a set of thresholds. All that is needed is to select appropriate levels for B_{OP} and B_{RP} thresholds, as shown in Figure 8-9. Because of variation in the magnetic baseline as measured over the operating airgap, it may not be possible to select a single set of thresholds that allows for operation over more than a small range of airgap. Operating the sensor closer than some minimum airgap will result in its output being stuck "on," while attempted operation beyond a maximum airgap will result in a permanent "off" condition. In addition, as one approaches either airgap extreme, output duty cycle can vary significantly, and the output waveform may not necessarily be a good representation of the target profile.

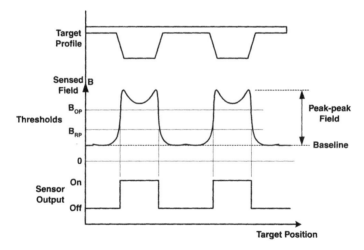

Figure 8-9: Thresholding a signal from a single-point sensor to obtain an output signal.

While conceptually simple, in practice this fixed-threshold scheme runs into serious difficulties. The biggest problems result from the high levels of baseline signal in relation to peak-to-peak signal. Flux variation in the sensor assembly's magnet, resulting from normal materials variation, will almost always require trimming the B_{OP} and B_{RP} points of each sensor produced. But even this will not eliminate baseline variation problems; temperature coefficients in the magnet material and the sensor can also cause baseline shifts over temperature that can't be easily trimmed out. Finally, it is difficult to obtain sensors that can be used at useful baseline levels. Most digital Hall-effect switches have B_{OP} points of less than 500 gauss, while most modern linear devices are limited to a sensing range of about ±1000 gauss before they saturate. The field strength at the face of an Alnico magnet can be 1000–1500 gauss, while that of a rare earth magnet can be in the range of 2500–5000 gauss. While one can shim the sensor off the face of the magnet to reduce the field to easily measured levels, this technique will also reduce the resulting sensor's maximum working airgap. Since few, if any, integrated switches or linear sensors operate at these high field levels, one may have to use discrete transducer elements and signal processing circuitry, which can result in increased system cost.

Because a high baseline field can make the design of an effective fixed-threshold sensor difficult, several schemes have been developed to remove baseline signal though clever magnetic design. Because the offset cancellations are performed magnetically within a single magnet, or multiple magnets of the same material, these techniques provide good temperature stability.

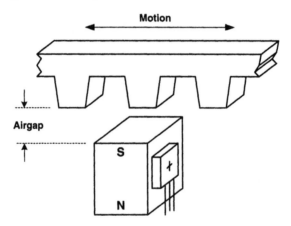

Figure 8-10: Lateral-field speed sensor.

One scheme [Wolfe90] relies on detecting the flux emanating from the side of the magnet. When no target is present, there is normally a null-point of zero net flux located halfway between the poles of the magnet (Figure 8-10). When a target approaches one of the magnet's poles, it causes a redistribution of flux in and around the magnet, and subsequently causes the null-point to move away from the magnet's center and closer to that pole. The sensor then experiences a nonzero flux, which can cause it to turn on.

Transducer placement is critical to the proper operation of this type of sensor, and is also application dependent. For optimum performance, sensors must be adjusted to match the target with which they are intended to operate.

Because the perturbation caused by an approaching target is spatially "filtered" through the bulk of the magnet, we would not expect as high a peak-to-peak response to small targets (as compared to the size of the magnet) as we would get where the sensor is placed between the magnet pole and the target. In many cases, however, this characteristic can be useful as it results in reduced sensitivity to sharp corners and surface roughness.

A second baseline offset removal scheme [VIG98] uses a compound magnet in the form of a "sandwich," as shown in Figure 8-11. In this arrangement, the normal flux from the inner magnet layer cancels the flux from the outer layers in the vicinity of the Hall-effect transducer, effectively creating a null point near the face of the compound magnet. An approaching ferrous target, however, causes a flux perturbation of a magnitude near what one would expect from a single magnet of comparable size and composition. Because the transducer element is at the "pole-face" of the compound magnet, in close proximity to the target, this architecture should yield good sensitivity to small target features. A commercial example of this type of sensor is the Allegro ATS535 geartooth sensor module. This device consists of a programmable switch over-molded into an assembly with a compound magnet. To implement a sensor using this assembly, one must present a target (often rotating) to the sensor assembly, and then electronically adjust the B_{OP} of the Hall effect IC so that it correctly senses the presence and absence of target features.

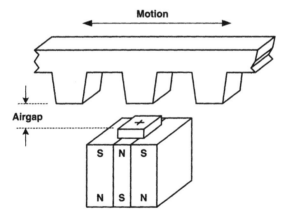

Figure 8-11: Speed sensor using magnetic "sandwich" to develop a null point.

8.8 Single-Point/Dynamic-Threshold Schemes

Because baseline flux variation makes it almost a necessity to adjust B_{OP}/B_{RP} points on individual sensors, single-point/fixed-threshold sensing schemes can be difficult to implement and use effectively. If the sensor were vested with the ability to adjust its operating thresholds to accommodate both target, magnet, and sensor variations, it would result in a more flexible and easy-to-use device.

The first, and oldest, dynamic threshold detection scheme is often called either DC blocking or AC coupling. A block diagram of this type of sensor is shown in Figure 8-12.

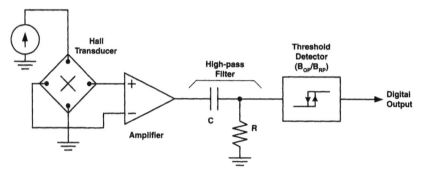

Figure 8-12: AC-coupled geartooth sensor.

The above circuit blocks any DC bias signal from getting through to the threshold detector. The resulting signal, as seen by the threshold detector, has an average over time of zero. The operate and release thresholds, therefore, are often set symmetrically about zero. Because the DC average is removed from the input signal, the sensor behaves as if it were adjusting its operate and release points to accommodate that average. Figure 8-13 shows idealized versions of some of the waveforms inside an AC-coupled sensor. The input signal is first level-shifted so that it is symmetric about a "zero" point. This signal is then compared to B_{OP} and B_{RP} levels that are commonly set to be symmetric about zero. This results in an output waveform that is independent of the baseline flux measured by the transducer.

While it is possible to base an effective geartooth sensor on an AC-coupled architecture, the scheme has a number of limitations. The first is that the target has to be moving at a certain minimum speed in order to actuate the sensor. Because the low-frequency response of the first-order RC filters typically used in this type of sensor rolls off gradually, one does not see a sharp minimum speed below which the sensor ceases to function. A more graceful degradation occurs, with the effective maximum airgap declining with target speed. Additionally, sharp features in the target introduce harmonics into the input signal. A square-wave signal with a frequency of 1 Hz also contains significant energy at 3 Hz and 5 Hz. These harmonics can also cause the sensor to switch. Because the harmonic content of the input signal is dependent on the airgap and geartooth geometry, the shape of the gear target also has a significant effect on lower operating frequency.

Another disadvantage of a simple AC-coupled scheme is that, in order to get a large effective airgap, one must set the operate and release points to a low value. While this results in a sensor that provides satisfactory performance at moderate to wide airgaps, overshoot effects can result in spurious output pulses when the target is too close to the sensor.

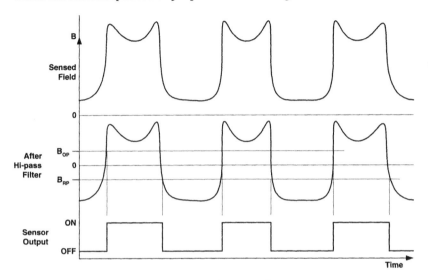

Figure 8-13: Waveforms in AC-coupled geartooth sensor.

Finally, an AC-coupled sensor will have a long "wake-up" time when it is first powered on. When power is first applied, several RC time constants will usually be required for the DC-blocking filter to stabilize. During this period the output of the sensor may be erroneous. For a "typical" AC-coupled geartooth sensor, with a low-frequency corner of approximately 5 Hz, this wake-up time can amount to as much as a few hundred milliseconds. For applications requiring instant wake-up on power-up, such as automobile ignition systems, this can be a serious drawback.

Another dynamic threshold detection scheme is related to AC coupling, but uses the capacitor to temporarily store the peak value of the signal [VIG95]. The peak value is then used to determine the values of the operate and release points. By deriving the B_{OP} and B_{RP} points from the peak flux, the sensor can adjust itself to the characteristics of both the magnet used to bias it, and the target being sensed. A block diagram of this technique is shown in Figure 8-14.

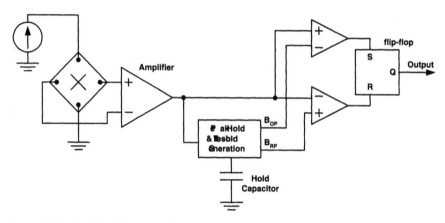

Figure 8-14: Peak-hold geartooth sensor.

This technique has a number of advantages over the AC-coupled detection scheme. The first is that a peak-detecting circuit will respond very quickly when establishing a peak value and hold that value for a substantial amount of time. This means that wake-up response can be very fast, with the circuit stabilizing as soon as a signal peak is encountered. Since the decay rate of a peak-hold circuit can be made long compared to the peak capture time, it is also possible to sense targets moving at very low speeds, at least compared to AC-coupling schemes. Finally, by scaling the values of the operate and release points to the magnitude of the peak incoming signal, this type of signal-processing system yields much better timing performance than the simple AC-coupled scheme.

It is also possible to perform signal-processing operations in the digital domain and, as integrated circuit processes yield higher component densities, this will become an increasingly attractive option. When one digitizes the transducer signal, it becomes possible to perform some very sophisticated processing to derive an output signal. The general block diagram of this type of signal-processing system is shown in Figure 8-15.

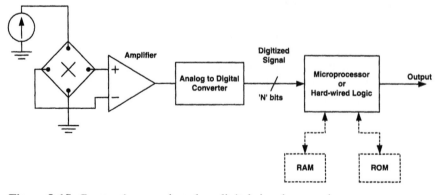

Figure 8-15: Geartooth sensor based on digital signal processing.

The use of DSP offers the potential of choosing from a great number of processing techniques. One example is found in the Melexis 90217 geartooth sensor. The amplified analog transducer signal is first converted into a digital form through an A/D converter. Digital logic then looks for the minimums and maximums in that signal. When a maximum is passed, and the signal declines by the hysteresis amount, the output of the sensor is turned ON. When the minimum is found and the signal increases by the hysteresis amount, the output will switch OFF. Because bits in a digital storage register are used to store temporary reference values, as opposed to representing these values as charge on an analog capacitor, this type of sensor can detect targets moving at arbitrarily low speeds.

8.9 Differential Geartooth Sensors

By placing two Hall-effect transducers close to each other on the same IC, it is possible to obtain an accurate approximation of the spatial gradient of magnetic flux density [Avery85]. On most commercially available differential geartooth sensor ICs, the Hall transducers are spaced about 2 mm apart. By subtracting the signals from the individual transducers, it is possible to derive an approximation to the gradient in the region near the sensor. When such a sensor is placed on the pole-face of a magnet, as shown in Figure 8-16a, a signal such as that shown in Figure 8-16b may be obtained in response to a passing gear.

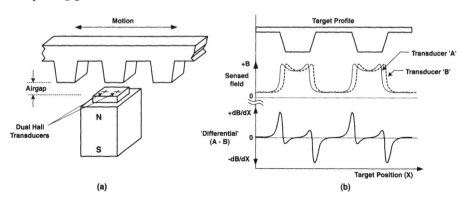

Figure 8-16: Positioning of differential transducer (a) and signal resulting from passing gear (b).

The first thing to note is that this sensor only provides a response to edges of target features. Flat features, be they gear teeth or the spaces in between, do not elicit any response. In the case shown above, a tooth to the right of the sensor results in a positive signal, while a tooth to the left of the sensor results in a negative signal. At all other times the signal is zero.

The edge-detecting behavior of a differential sensor offers a number of advantages over single-point sensing schemes. The primary advantage of an "ideal" differential signal is that, because one is looking for positive and negative signal events against a zero-level background, it is easy to identify target edges. Differential sensing should also make it easier to build a sensor whose performance is less susceptible to sensor, target, and airgap variations than that of a single-point sensor. While AC-coupling or other variable-threshold techniques can be used to detect edges by the signal's change in time, a DC-coupled differential sensor can detect an edge even when the target is standing still.

Another potential advantage of a differential geartooth sensor is that of more accurate timing response. Because one is looking at the spatial flux gradient caused by the edge of a target feature, it is easier to locate the edge accurately with differential sensing methods than with single-point ones.

Despite their potential advantages, differential sensing schemes also have their share of quirks. Two of these are orientation sensitivity and phase-reversal.

To sense an edge, a differential sensor must be oriented so that the sensor IC is straddling that edge, with one of the Hall-effect transducer elements in proximity to the geartooth, and the other in proximity to the space between teeth. If the sensor is rotated 90°, however, so that both transducers "see" the tooth or both transducers "see" the gap, then it will not detect a gradient. While 90° of sensor rotation represents the worst case (no signal out), performance will degrade as the sensor is rotated from its optimal working position to the point where it ceases to detect the target.

If one rotates the sensor 180° from normal working position, the sensor will still output a signal; the polarity, however, will also flip. If leading edges were represented by positive output signals and trailing edges by negative output signals when the sensor was correctly oriented, the leading edges will be represented by negative signals and the trailing edges by positive signals if the sensor is rotated 180°. While this effect may not be an issue in speed-sensing applications, where all one cares about is the total number of pulses, it can cause problems in systems that are specifically looking for particular edge transitions. One example of an application in which mounting a differential sensor backwards could cause havoc is the timing sensor for an automotive ignition system. Reversing the polarity of the output train prevents the engine controller from accurately identifying critical timing angles, and consequently from running the engine properly.

Another counter-intuitive behavior unique to differential sensors is polarity inversion resulting from the direction in which the target is rotating. With a differential sensor, if you rotate the target backwards, the phase of the output signal inverts, in a manner similar to that described above. There are situations, however, in which this effect can be put to good use. If you make a target that results in an output signal with a duty cycle that is not 50% (output is HIGH 50% of the time), you can determine the target's direction of rotation. For example, if the target is cut so that the duty cycle in the forward direction is 75%, you will see a 25% duty cycle when the target is rotated backwards.

If differential geartooth sensors exhibit all of the strange behaviors described above, why do people use them? The reason is that, when applied properly, they offer good timing accuracy and generally high performance. And, as is the case with single-point sensors, there are numerous signal-processing strategies that can be used to optimize performance for a given set of applications.

8.10 Differential Fixed-Threshold

The simplest signal-processing method for differential geartooth sensors is to set a fixed threshold. Because it is necessary to look for both positive and negative signal peaks, however, a pair of thresholds is required. Ideally, these thresholds will be set symmetrically about zero, such that $B_{OP} > 0$, $B_{RP} < 0$, and $B_{RP} = -B_{OP}$. By remembering the polarity of the last "edge event" the output of the threshold detector will track the target profile. The block diagram in Figure 8-17 shows the major functions and signal flow of a differential fixed-threshold speed sensor.

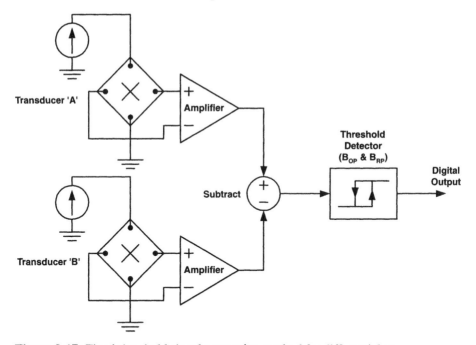

Figure 8-17: Fixed-threshold signal processing method for differential sensor.

While this method can be made to work, there are numerous issues to be addressed to ensure a successful implementation. The first, and most important, is that of offset error. The signal seen by the threshold detectors can have a constant error signal added. If this

error signal is large enough compared to the threshold levels, the sensor may sporadi-
cally or completely malfunction. Some sources of offset error are:

- Electrical offsets in the transducers and front-end instrumentation amplifiers
- Nonuniformity in the magnetic field provided by the bias magnet
- Tilting of the sensor in relation to the target.

While these offsets can be minimized through appropriate design techniques, both
at the IC and assembly level, they pose a limit to how low the thresholds can be set, and
consequently limit the maximum working airgap of the assembly. For these reasons,
while fixed-threshold differential sensors can be used as the basis of practical sensor
assemblies, they are not usually the best choice for most applications.

8.11   Differential Variable-Threshold

As was the case with single-point geartooth sensors, it is also possible to use variable-
threshold signal-processing schemes with differential geartooth sensors. Two of the
ways in which thresholds can be adjusted are by shifting them (to reduce the effects of
offset) and by widening or narrowing them (to adapt to changes in signal magnitude).

Shifting the thresholds or shifting the sensed signal can be used to reduce the effects of offset. A signal-processing scheme that performs this function will allow the
geartooth sensor to cope with variations in the transducers, magnets, and alignment of
the sensor assembly. One of the simplest circuits used to perform this level-shifting operation is the AC coupler that was presented for use with single-point geartooth sensors.
A detailed description of this type of sensor can be found in [RAMS91]. A representative device of this type is the Allegro UGN3059.

While shifting the input signal so as to remove offset solves many problems and
results in a sensor with greatly increased performance and ease-of-use over that of a
fixed-threshold device, it is not a panacea. First of all, the AC-coupling scheme places
a lower limit on detectable target speed, just as it did in the case of a single-ended
AC-coupled sensor. Additionally, a single set of symmetric thresholds doesn't provide
optimal performance under all operating conditions. If the thresholds are set high, then
the sensor's maximum working airgap is limited. If one wants to increase the maximum
airgap by setting the thresholds very low, false triggering from target corner effects and
surface roughness can occur.

One solution is to adjust the switching thresholds based on the magnitude of the
incoming signal, as was shown in the case of the single-point sensor. If the magnitude of the incoming signal is large, then the thresholds can be set high. If the signal
magnitude is low, then the thresholds also become low. The signal magnitude may be
determined either by taking its absolute value and averaging, or by detecting its peak
value and holding that. By adjusting the thresholds to match the signal, a wide working
airgap range can be achieved. An additional benefit of this approach is that it will also

tend to increase accuracy in the degree to which the sensor's signal output tracks the features of the target, a useful feature for timing-sensitive applications. An example of a commercially available sensor that uses this type of signal-processing technique is the Allegro ATS610LSB geartooth sensor module.

Another way to achieve this effect is to keep the thresholds constant, but to provide a variable-gain amplifier whose gain is inversely proportional to the average signal magnitude. This type of compensation method is referred to as automatic gain control, or AGC. An example of a commercially available speed sensor that works in this manner is the Allegro ATS612 module.

8.12 Comparison of Hall-Effect Speed Sensing Methods

Because of the wide range of applications in which geartooth sensors are used, there is no single solution that works well everywhere. Table 8-2 summarizes a few of the advantages and drawbacks of the various geartooth sensing schemes described above. Examples of sensor ICs employing each of these techniques are also provided.

Table 8-2: Comparison of various geartooth sensing techniques.

Technique	Advantages	Drawbacks	Representative Devices
Single-Ended Fixed-Threshold	Power-on recognition True zero-speed sensing Orientation insensitive	Limited operating airgap range Needs to be adjusted to target and magnetics, often on a unit-unit basis Fair edge timing accuracy	A3250 (Allegro) Also many switches and latches with appropriate magnetics
Single-Ended Variable Threshold	Near-zero speed sensing Orientation insensitive Easy to Use	No power-up recognition Minimum sensing speed (in some cases)	ATS632 (Allegro) MLX90217 (Melexis)
Differential Fixed Threshold	Zero-speed sensing Good edge timing accuracy	Limited airgap range Some nonintuitive behavior	UGN3056 (Allegro) HAL300 (Micronas)
Differential Variable Threshold	Good edge timing accuracy Wide effective airgap range Easy to use	Minimum sensing speed (in some cases) Some nonintuitive behavior	UGS3059 (Allegro) ATS610 (Allegro) TLE4921-3U (Infineon)

8.13 Speed and Direction Sensing

One previously mentioned feature of differential speed sensors is that, with an appropriate target, they can also be used to determine the direction of rotation. If one knows the direction in which a target is rotating, it is possible to track the angular position of the target. Accurately measuring the angular position of a shaft is a very useful thing to be able to do, and is the basis for many types of precision positioning equipment.

The most common way of determining rotational direction is to use two separate sensors, sensing the same target at slightly different points, as shown in Figure 8-18a. To make for simple and reliable target detection, the target in this case is a ring magnet, and the sensors are digital Hall-effect latches. By spacing the sensors half a pole spacing apart, two output signals are obtained which are 90° out of phase with each other, with the leading edge of one coming either before or after the leading edge of the other. The direction of target rotation determines which signal will lead, and which one will lag. In the case of the waveforms shown in Figure 8-18b, clockwise rotation causes the A channel to lead while counterclockwise rotation causes the B channel to lead. This quality in a pair of signals is often referred to as quadrature.

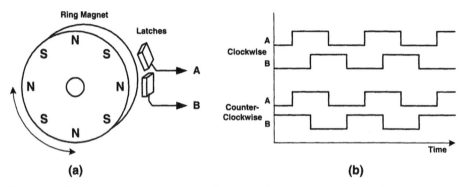

Figure 8-18: Using two sensors (a) to obtain quadrature output signals (b).

To determine rotational direction requires a circuit that can determine the lead-lag relationship between the two signals. One of the simplest circuits that can do this consists of a single D-type flip-flop (Figure 8-19a). The B channel signal is fed into the clock input, while the 'A' channel signal is fed into the D (data) input. The way a D-type flip-flop works is that, whenever there is a rising edge on the clock input, the flip-flop will instantaneously sample the state of the D input. It will then update its output (Q) to that sampled state and hold it until the clock input sees another rising edge. Figure 8-19b shows some sample waveforms resulting from a target that is moving forward, then reverses direction.

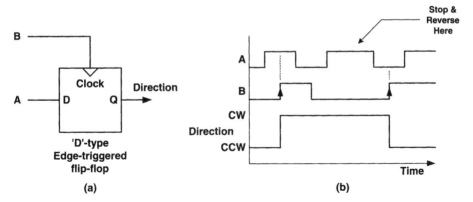

Figure 8-19: Using a D flip-flop (a) to determine direction from quadrature signals (b).

A direction signal can be used in conjunction with one of the sensor signals to track relative motion. The direction signal is used to determine whether a counter will increment (forward direction) or decrement (reverse direction) on receipt of a pulse from one of the sensors. The total count will represent relative motion from the time at which it was last reset. In practice, somewhat more complex logic than that described here is typically used to ensure the accuracy of the count, and therefore reduce position-tracking error.

While it is possible to implement a "speed and direction sensor" from a pair of digital Hall-effect latches and a D flip-flop, it is also possible to obtain this function in the form of a single integrated circuit. The Allegro Microsystems A3421 and A3422 ICs provide the necessary dual sensors, spaced 1.5 mm apart, as well as all the logic necessary to develop both direction and count output signals. For those situations in which one needs only the two sensor output signals (A and B) the Melexis MLX90224 omits the quadrature-decode logic and provides the digital quadrature signals only.

8.14 How Fast Do Speed Sensors Go?

One question that comes up frequently in the context of geartooth sensor performance is that of how fast a target can move and still reliably be detected by a particular geartooth sensor. The answer is quite complex and depends on a number of factors.

First, let's define "speed" in terms of the number of passing targets-per-second that the sensor can detect without error (adding or losing output pulses). This lets us define maximum operating speed in hertz. As an example, a 25-tooth target moving at 6000 RPM would present 2500 teeth per second to the sensor, equivalent to an operating frequency of 2.5 kHz. Let us also assume that the sensor is operating with a suitable target at an airgap well within its operating range, and therefore has a strong magnetic signal from which to operate. A sensor that operates marginally at low speeds will probably fail at slightly higher ones.

The first limitation is within the sensor IC itself. While the frequency response of the Hall transducer itself is very high (well in the megahertz range for those used on Hall ICs), the signal-processing circuitry can impose additional limitations. Chopper-stabilization circuitry can limit the IC's bandwidth to a few tens of kilohertz. Many signal-processing systems have bandwidths that extend beyond several hundred kilo-hertz. The exact frequency response of the sensor IC is dependent on the design and the process technology with which it was implemented. Many (but not all) Hall-effect sensor ICs, however, are capable of accurately sensing targets at frequencies in excess of 25 kHz. If this doesn't sound very fast, remember that we are discussing mechanical systems. A 25-kHz output signal corresponds to a 25-tooth target rotating at 60,000 RPM.

In many cases, however, the silicon is not the limiting factor when it comes to maximum target-sensing speed. The techniques used to package the sensor assembly, and even the IC, can introduce significant speed limitations.

While objects made from nonferrous metals, such as brass or aluminum, don't have many significant interactions with steady-state DC magnetic fields, they can in-teract strongly with time-varying ones. This is because a time-varying field tends to set up eddy-currents in any conductive body inside the field. These eddy currents flow in a manner that tries to prevent an externally applied magnetic field from entering the conductive body. Conversely, once a magnetic field is established inside a conductive body, eddy currents will flow in a manner that tries to prevent the field from leaving. Qualitatively, this effect is illustrated in Figure 8-20.

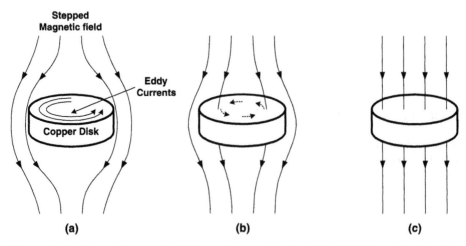

Figure 8-20: Stepped magnetic field entering conductive disk. Field immediately after field turns on (a), a few microseconds later (b), steady-state condition (c).

When an external magnetic field is initially applied to a conductive disk as a time-step function, eddy currents formed in the disk will try to exclude the external field (Figure 8-20a). After a short amount of time, these eddy currents will begin to die out, and the field will begin to enter the disk (Figure 8-20b). Finally, after some additional time, the eddy currents will have diminished to insignificant values, and the steady-state field will pass through the disk as if it weren't there (Figure 8-20c).

This effect occurs to varying degrees in all metal parts of a sensor housing through which sensed flux must travel. In general, the larger and thicker the metal body, the more it will attenuate time-varying fields. Also, materials with high electrical conductivities, such as copper and aluminum, will exhibit more pronounced effects than those with lower conductivities, such as bronze or zinc. This effect even occurs to some extent in the leadframes used in the Hall-effect sensor IC packages, although the time constants involved for structures this small can be measured in microseconds.

Since these dynamic eddy current effects are very complex to model analytically (or even with finite element analysis), it is difficult, if not impossible, to quantitatively predict their effects on sensor performance. Being aware of their existence, however, can provide at least a little insight that may be useful in helping to design sensors that meet your performance goals.

Chapter 9

Application-Specific Sensors

Many applications require the use of some support electronics in addition to a digital or linear type of sensor. If an application presents a sufficiently high manufacturing volume for a sensor requiring auxiliary electronics, it is often cost-effective to integrate those electronics onto the IC along with the sensor. An integrated circuit targeted towards a particular application is usually referred to as an application-specific integrated circuit, or simply an ASIC. The geartooth sensors described in Chapter 8 are examples of ASICs designed to sense spinning targets. The ICs described in this chapter are all designed to solve particular and unique problems.

9.1 Micropower Switches

User-interface controls, such as buttons and switches, increasingly control logic-level signals as opposed to directly controlling high-power loads such as motors. In low-current logic-level applications, however, mechanical switches exhibit surprisingly low reliability. Switching-current levels ranging from microamperes to milliamperes do not cause enough arcing to clean oxides from switch contacts. Additionally, logic-level voltages (3–5V) are not especially effective at breaking down oxide layers on switch closure. As a result, a mechanical switch that performs well when directly switching a power load may fail prematurely when used as a logic-level control.

Hall-effect sensors, in combination with an actuator magnet, can be used to make high-reliability user-interface controls. Since there are no contacts to corrode or wear, and no moving parts other than the actuator magnet assembly, it is possible to build logic-level controls with high expected mean-time-to-failure.

One drawback of using Hall-effect devices, however, is their relatively high current drain. A mechanical switch requires essentially zero current when not switched on. A Hall-effect switch may require 5 mA. While this may not be an issue if the control is going into a piece of line-powered equipment, for battery-powered devices low current drain is of paramount importance. For example, a typical 9V alkaline battery may be rated at about 500 mA-hours, meaning that it can deliver 1 mA for 500 hours (or 10 mA for 50 hours) before being drained. A Hall-effect sensor consuming 5 mA could only be operated for about 100 hours from this battery, or just a little over 4 days. This might not be a problem if these 100 hours represented "usage" time for the device. If the Hall-effect sensor, however, is the ON/OFF switch for the device, having to change the batteries every 4 days, whether or not the device is actually being used, would be a real problem. A battery-replacement schedule this frequent would almost certainly make the device impractical as a product.

One solution to the problem of current drain was presented in Chapter 5, which is to power the device only when you need to look at the output. While only a small amount of electronics is needed to perform the actual power switching, the electronics necessary to control the power-cycling and retain the output state of the Hall-effect sensor can be considerably more costly. The additional cost to power-cycle the sensor will often make this solution prohibitively expensive for many consumer applications.

Because portable battery-operated applications are becoming increasingly popular, a significant market has emerged for low-power Hall-effect sensors. For this reason, several manufacturers now offer micropower Hall-effect devices. These devices operate by power-cycling the Hall-effect transducer and its associated bias and amplifier circuitry. Figure 9-1 outlines how this is done. A low duty-cycle pulse generator turns on the transducer, amplifier and threshold-sensing circuitry. On the falling edge of the pulse, the outputs of the threshold sensors are latched.

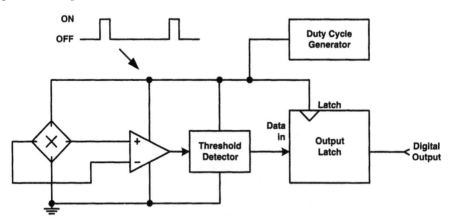

Figure 9-1: Schematic representation of power-cycled Hall sensor.

The Melexis MLX90222 is one example of a micropower Hall-effect switch. In this device, the Hall-effect transducer and amplifier are powered ON for 60 μsec (typical), during which a flip-flop is either set or reset based on whether the sensed magnetic field is greater than the device's magnetic operate point (B_{OP}) or lower than the release point (B_{RP}). In this ON state, the sensor IC draws about 0.7 mA of supply current. At the end of the ON state, the transducer and amplifier are then turned off for 40 ms (typical), after which the cycle repeats. While the transducer is in a powered-down state, the output state is maintained by a flip-flop, and the supply current is only about 10 µA. Because the total fraction of time spent in the ON state is only about 0.15%, the average current drain is only slightly higher that of the powered-down state.

Because the output state cannot be affected by changes in magnetic field when the sensor is powered down, the overall effect is like having the sensor take a "snapshot" of the applied magnetic field about 25 times a second. Another way of expressing this is that the sensor has a 25-Hz sampling rate. This is sufficiently fast for many types of user-interface controls. For a few user-interface devices, however, such as computer keyboards, a 40-ms response time may be sufficiently long to change the overall "feel" of the interface.

One application where a 25-Hz sampling rate would probably be unacceptable is where the device is used in conjunction with a ring magnet as a speed sensor. It is not difficult to envision situations where more than 25 poles/second are presented to the sensor. Even in cases where a few discrete magnets are used, the low sampling rate makes it easy for a magnet to "sneak by" the sensor undetected.

The Allegro A3210 also incorporates power-cycling circuitry, but includes an additional feature that makes it very easy to use as a replacement for magnetic reed switches. Where most Hall-effect switches turn on in response to a particular pole of a magnet, usually the south pole, the A3210 will switch ON in response to either a north or a south pole. This type of omnipolar behavior (the term "bipolar" is already taken) means that it doesn't matter which way you put the magnet into an assembly, since the sensor will turn on when either magnet pole approaches it. The magnetic response curve for this device is shown in Figure 9-2.

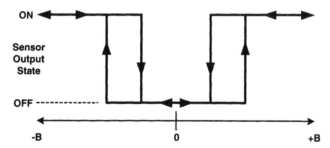

Figure 9-2: Hysteresis curve for omnipolar A3210.

The A3210's omnipolar behavior is extremely valuable when using the devices in high-volume manufacturing environments, because it makes the assembly insensitive to magnet orientation. Many assemblies will use rod or bar magnets that are mechanically symmetric with respect to the poles, meaning they can fit into a larger assembly in two orientations. With traditional south-pole sensitive Hall-effect switches, one of these possible orientations will result in a nonfunctional assembly. Because the A3210 will respond to either pole, a bar magnet can be inserted into the assembly in either north-facing or south-facing orientations and the assembly will still function.

9.2   Two-Wire Switches

Transmitting data signals in the form of a current as opposed to a voltage often results in a higher immunity to external electrical noise sources. For this reason, signaling methods such as the "4–20 mA current loop" are popular in noisy environments such as factory floors. Another advantage of current-loop signaling is that if the sensor can be powered by a voltage applied across the transmission leads, only two wires (V+, V–) are required, as opposed to the VCC, GROUND, OUTPUT triad required by voltage-signaling schemes.

Both because of enhanced noise immunity, and the economies offered by saving a wire, numerous automotive applications are moving to two-wire signaling schemes. Consequently, several Hall-effect sensor vendors are now providing integrated two-wire Hall-effect switches.

While one can convert a standard three-wire type of Hall switch into a two-wire device through the addition of external circuitry, there are various drawbacks to this approach. In the simplest case, a resistor may be tied between VCC and OUTPUT, as shown in Figure 9-3. When the device switches ON, the resistor is pulled to ground by the device's output, and draws additional current. The magnitude of this on-state current, however, will vary linearly with the supply voltage. While additional circuitry can be added to make the on-state current less supply-dependent, the additional components can get expensive, and this also has no effect over the off-state current, which will be dependent on the quiescent current of the sensor.

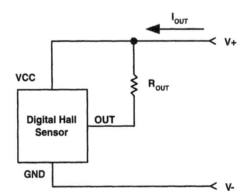

Figure 9-3: Converting a three-wire Hall-effect switch into a two-wire switch.

Two-wire integrated Hall sensors, on the other hand, provide well-specified on-state and off-state currents. This makes it much easier to design interface circuits that can reliably read their state despite power supply, temperature and manufacturing variation. The Melexis 90223 is an example of a two-wire Hall-effect switch. A functional block-diagram representation of this part is shown in Figure 9-4.

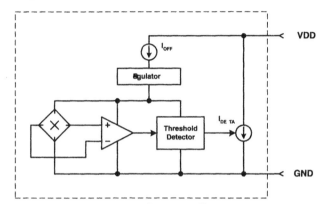

Figure 9-4: Melexis MLX90223 two-wire Hall-effect switch.

When no field is applied to the device and it is in the on-state, the current drawn through the VCC terminal ranges from 3.9 to 6.9 mA. When the applied magnetic field exceeds B_{OP} (60 gauss typical), the output current source is switched on, increasing the current drawn from the VCC terminal to between 11 and 19.4 mA.

A simple circuit for interfacing to a MLX90223 is shown in Figure 9-5. The output of this circuit is compatible with TTL or CMOS logic inputs. A 51Ω resistor is used to convert the supply current to a voltage. This voltage is then fed to a comparator, where it is compared to a threshold voltage equivalent to 9 mA of current (9 mA × 51Ω = 459 mV). This represents a value halfway between the maximum off-state current level (6.9 mA) and the minimum on-state current level (11 ma).

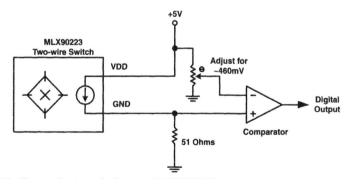

Figure 9-5: Circuit for interfacing to a MLX90223 two-wire switch.

In an actual application and depending on the operating environment, one may wish to add additional circuitry to provide noise filtering and protection from electromagnetic interference (EMI) and electrostatic discharges (ESD). A small amount of hysteresis added to the comparator will reduce the degree to which false transitions occur at its output (due to various noise sources). Finally, with some modifications, mostly in the voltage supplied to the sensor and the level of the HIGH/LOW threshold, this circuit can be adapted for use with other two-wire sensors. Because two-wire systems are becoming increasingly popular, several Hall-effect IC suppliers now offer two-wire sensors. Table 9-1 shows a few of these offerings:

Table 9-1: Two-wire Hall-effect switches.

Supplier	Two-wire Switch Models
Allegro Microsystems	A3161, A3361, A3362
Melexis USA	MLX90223
Micronas	HAL556, HAL566

9.3 Power Devices

While methods of driving incandescent lamps and relays were outlined in Chapter 5, these approaches all require the use of external discrete components. While this is acceptable in many circumstances, the use of external components can result in a design that is both physically large and relatively expensive. The additional solder joints required can significantly reduce the reliability of a design.

The Allegro UGN5140 high-current Hall-effect switch provides a single-chip means of driving small lamp and electromechanical loads. It offers an open-collector output capable of sinking up to 600 mA. It also provides an integral fly-back diode, for use when driving inductive loads such as relays and solenoids. Figure 9-6 shows a schematic representation of the UGN5140.

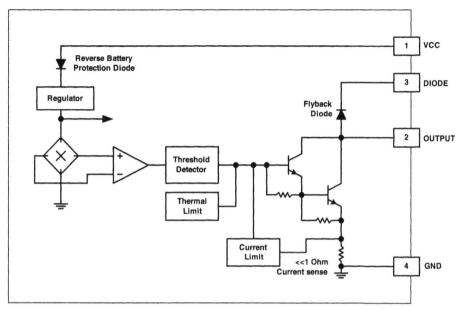

Figure 9-6: UGN5140 High-current Hall-effect switch.

The UGN5140 also provides two features that would be relatively difficult and costly to provide in a discrete power-driver. The first of these is output current limiting. In the event of a short-circuit or similar over-current condition on the output, the device will limit its output sink current to a preset threshold (typically 900 mA). This is useful when driving incandescent lamp loads (potentially several amperes), as it prevents the lamp cold-filament inrush current from damaging the sensor. The second feature is a thermal shutdown. If the internal die temperature of the UGN5140 exceeds 165°C (nominal), the sensor's output shuts off. Combined with output current limiting, this feature is useful in enabling the sensor to survive various types of output fault conditions, such as short circuits.

9.4 Power + Brains = Smart Motor Control

The Melexis US79 is an extreme example of an application-specific Hall IC in that it was designed to perform only one job: to provide control and power-driver functions for a small 2-winding brushless DC motor. This type of motor is commonly used in the cooling fans found in virtually all personal computers. A typical DC brushless fan, disassembled to expose the motor, Hall sensor, and drive electronics, is shown in Figure 9-7.

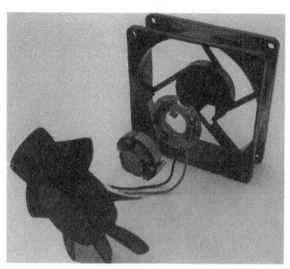

Figure 9-7: DC brushless fan using Hall-effect sensor (photo courtesy of Melexis USA).

Brushless DC motors operate by using several fixed windings to generate magnetic fields that move a magnet embedded in the rotor, causing the rotor to spin. To keep rotation continuous, the windings must be energized and de-energized at particular times depending on the position of the rotor. One way to do this is to use a Hall-effect sensor to detect the position of the rotor magnet. This information is then used to determine which winding to energize at any given time. This process of switching the windings on and off to make the motor spin is called *commutation*. Figure 9-8 shows a schematic of how the US79 is used to drive the two windings of a brushless fan. No other electronic components are required.

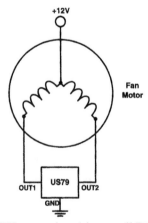

Figure 9-8: Schematic of US79 as used to drive small DC brushless motor.

Traditionally, in a brushless DC fan motor, commutation is performed with a Hall-effect sensor and numerous discrete components. While this approach works reasonably well, there are several areas for potential improvements, which are addressed by the US79. The first is component count reduction. While the cost of the individual components used in a discrete design may not be very much, for a high-volume item such as a cooling fan, even pennies count. Each discrete component used also adds assembly cost. Assembly costs in some cases can actually exceed the costs of the components. By replacing from 8 to 15 discrete components with a single device, the US79 can make the design more cost-effective.

A second advantage of a highly integrated solution like the US79 is increased reliability. Reducing component package count and the number of solder connections can greatly increase a system's mean-time-between-failures (MTBF). In an application like a computer cooling fan, the failure of the fan can cause subsequent additional failures of other components through overheating. So although the fan itself may be inexpensive and easily replaceable, subsequent failures of other parts may be considerably more costly, imposing high reliability requirements on the fan.

Finally, another advantage offered by an integrated solution over a discrete one is that significant intelligence can be placed on silicon in an IC at very little incremental cost. In the case of the US79, this includes the ability to recognize and appropriately respond to various fault conditions. For a DC fan, one of the most common failure modes is for the rotor to become mechanically jammed. This state is referred to as a *locked rotor* condition. If the commutation circuit keeps applying continuous power to a locked rotor, potentially hazardous overheating can occur, either in the motor windings or in the driver circuit. By being able to recognize a locked rotor condition, the US79 is able keep the fan from overheating. Additionally, the US79 will periodically try to restart the fan, so that if the fan rotor becomes freed, the fan will resume normal operation. Performing these functions with a discrete circuit would be complex and probably not economically feasible to implement in a device as cost-sensitive as a computer cooling fan.

This chapter has described a few of the application-specific Hall-effect sensors that are commercially available. As new applications are developed, there are sure to be many more new types of these devices offered in the future.

Chapter 10

Development Tools

Sharp tools suited to the task at hand can make any job easier. The development of a magnetic sensor assembly is no exception. This chapter will briefly describe some of the tools that can come in handy when designing magnetic sensors. Some of these are common to any well-equipped electronics laboratory, while others are unique to the magnetics world. This chapter will describe what kinds of equipment I have found useful in the past for this type of work.

10.1   Electronic Bench Equipment

For most sensor development projects, a well-equipped electronics lab can make development easier, faster and more straightforward. While one can spend large amounts of money on the highest-performance, leading-edge test equipment, it is also possible to get the measurement capabilities one needs for a modest price.

Power supplies

In the course of developing Hall-effect sensor assemblies, I have needed power supplies for three purposes:

- Powering sensors and related circuits
- Providing precision voltage and current references
- Driving coils to generate magnetic fields

These applications all place different demands on a power supply. Although it may be possible to get a single unit that can perform adequately in all of these tasks, it is usually better and less expensive to get several, more specialized units.

For powering sensors and related circuits, such as laboratory breadboards, the typical requirements of a power supply are that it be able to provide modest (< 500 mA) amounts of current over the operating voltage range of the sensor (0–30V). An important feature to look for is user-settable current limiting. While most good power supplies have internal current limiting designed to protect the power supply from output short-circuit conditions, this internal current limiting won't usually protect your circuits from the power supply if they are miswired or otherwise misbehave. The ability to set power supply current limits to some arbitrarily low and nondestructive value is extremely valuable in protecting your work as you debug it.

Multiple outputs are also a nice feature; being able to get two or three separate voltages out of the same box can greatly reduce benchtop clutter. Many power supplies also offer built-in digital or analog front-panel meters, so you can monitor the output voltage and current. The accuracy of these built-in meters tends to be limited, so for critical measurements you may want to do the monitoring with a separate voltmeter or ammeter.

A precision power supply is quite different from a general-purpose bench supply. Many of these units will have digital front-panel controls, and can often be controlled externally through a IEEE-488 GPIB port or serial port. They can be programmed to either supply a voltage and measure the current drawn, or to supply a current and measure the voltage, often with five or more digits of accuracy. In contrast to the meters built in to conventional bench supplies, the meters on precision power supplies generally are quite accurate. These units are commonly used in automatic test equipment (ATE) systems. Agilent (formerly Hewlett-Packard) and Keithley Instruments are two of the more well-known manufacturers of these devices.

The last application for power supplies is for driving coils or electromagnets to produce magnetic fields. These applications can require significant amounts of current (> 10A) at moderate voltages (50V). Needless to say, the exact requirements are crucially dependent on the magnet or coil to be driven. One common requirement, however, is that the power supply be able to go into a constant-current mode. Driving an electromagnet with a constant current as opposed to a constant voltage reduces the resultant magnetic field's temperature dependence as the coil resistance increases from heating.

Voltmeters and DMMs

My favored approach here is to get a few handheld digital multimeters (DMMs) of moderate resolution (3½ digits), and one or two high-resolution (5–6 digit resolution) bench-top units. It is worth having several handheld DMMs around because there will be times when you may need two or three (or even more) for a bench setup. For most measurements, a 3½ digit handheld DMM will provide sufficient measurement accuracy and resolution; the 6-digit bench DMM is for those rarer situations where higher accuracy is really needed.

Although it is possible to buy very inexpensive handheld DMMs from a number of sources, it is my opinion that it is usually worth spending a few extra dollars to get higher quality instruments, for several reasons. First, the better instruments tend to be more rugged and last longer. Additionally, it is usually easy to obtain calibration services for name-brand instruments. Aside from the general desirabilty of having instruments whose readings have some correlation with reality, calibration can become a major issue if you or your shop are subject to various quality systems (e.g., ISO9000).

Oscilloscope

Because most sensors detect mechanical motion, the resulting electrical outputs tend to vary slowly, at least by electronic standards. If speed and bandwidth were the only considerations for the selection of an oscilloscope, then just about any oscilloscope made since World War II would probably be adequate for most Hall-effect sensor development work. Small, no-frills analog oscilloscopes can be obtained for a few hundred dollars. One characteristic of many sensor applications, however, is that one often needs to look at events that occur either rarely or on a single-shot basis. In these cases, a digital storage oscilloscope (DSO) is much more useful than a traditional analog one. Fortunately, the prices on digital oscilloscopes have fallen dramatically in the past few years, and it is now possible to get a good entry-level scope with two input channels and 60 MHz of bandwidth for less than $1500.

Frequency Counter

When working with rotating targets, a frequency counter is often useful to accurately determine target speed (RPM). Increasingly, however, this function is incorporated into DMMs and digital oscilloscopes, so it may be unnecessary to buy a separate instrument to obtain this measurement capability.

Clamp-On Current Probes

If you are developing current sensors, some means of measuring large currents is necessary. Most handheld DMMs only offer ranges up to about 10A. Additionally, to measure current with a DMM, you need to break into the circuit carrying the current. To measure larger currents noninvasively, clamp-on current probes can be used. These devices come in two fundamental varieties; AC and DC. AC current probes are based on inductive sensing principles, and will only measure AC current, while DC current probes can be used to measure both AC and DC currents. A DC clamp-on current probe works much like the current sensors described in Chapter 7; the main difference is that the magnetic path is set up so it can be readily opened and closed around a conductor. This allows one to make current measurements without having to break the current path. Many current probes don't have an integral display, but need to be connected to a DMM to be used.

Solderless Breadboard

One final and very useful piece of electronic equipment is the solderless breadboard. This device consists of a plastic block with lots of holes into which you can insert components such as resistors and DIP ICs. You then temporarily wire the components together by poking wires into nearby holes. Solderless breadboards allow you to quickly build prototype circuits and allow for easy changes and debugging. Breadboards are available in several sizes and configurations. Some of them also provide power supplies.

It should be pointed out, however, that breadboards are intended for prototyping circuits; any circuit you expect to keep around for a while should probably be constructed with other, more permanent techniques. Other than lack of permanence, circuits constructed on breadboards also have two other major limitations. The first is that the breadboards are designed to accommodate through-hole components. While it is possible to buy adapter bards to make surface-mount parts fit into a breadboard, this can be a nuisance. For designs using surface-mount parts you are usually better off designing circuit boards for your prototyping work.

The second major limitation of solderless breadboards is that circuits constructed on them typically have lots of parasitic capacitance and inductance. Although most Hall-effect sensor circuits will operate at audio frequencies (10 Hz–20 kHz), these parasitics can have adverse effects on other components that may have much higher frequency responses. For this reason, circuits built on solderless breadboards will often have somewhat different performance characteristics than the final circuit will when built on a printed circuit board.

Because of their utility and low cost, and despite their limitations, it is usually worth getting a few solderless breadboards to have around the lab.

10.2   Magnetic Instrumentation

While the equipment described in the last section can be found in nearly any electronics lab, some more specialized items can be useful when developing magnetic sensors.

Gaussmeter

A gaussmeter measures magnetic flux density (B) at a given point in space. Most gaussmeters employ Hall-effect sensor elements as the magnetic probe. In its simplest form, a gaussmeter is a linear Hall-effect sensor with a meter readout. Indeed, it is possible to build a simple gaussmeter from a linear Hall-effect sensor IC, a small amount of interface electronics, and a DMM, but the result would not provide anywhere near the capabilities of a modern gaussmeter. A few of the features to look for in a gaussmeter are:

- Range – How small a field can it measure, and how large a field can it measure?
- Accuracy – To what degree does the reading reflect reality?
- Interface options – In addition to a front-panel display, can it communicate with PCs or other instruments?

Range is important because there are times when you will want to measure fields of a few gauss, and others where you will want to measure fields of several kilogauss. Low ranges are often important in sensor work. Even though most Hall-effect sensor ICs aren't useful for discriminating field differences much below 1 gauss, you will typically want an instrument with an order of magnitude finer resolution than what you need to measure.

The need for accuracy requires little if any elaboration. Inaccurate instruments can make your life vastly more difficult. Accurate instruments, regularly calibrated, can make development work go more smoothly, by reducing one potential source of errors. Note that accuracy is a key specification for gaussmeters and is often the only difference between two instrument models of differing price.

While interface options may not seem that important, they enable one to hook the gaussmeter to a PC and automate many simple tasks. Popular interface standards include RS-232, IEEE-488, and analog outputs.

Fluxmeter

A fluxmeter measures changes in magnetic flux, as detected through a Helmholtz or similar coil arrangement. Functionally, a fluxmeter consists of a pickup coil and an electronic integrator, as shown in Figure 10-1. A change in total flux through the pickup coil induces a small voltage, which is then integrated over time. By integrating the voltage developed by the coil, which itself is proportional to the derivative of the flux passing through the coil, a fluxmeter can measure net changes in that flux. A fluxmeter is therefore different from a gaussmeter in two major ways. First, the fluxmeter measures total flux (ϕ) over an area, where the gaussmeter measures flux density (B) at a single, small, point in space. The second major difference is that, while a gaussmeter can resolve a zero flux condition (to some degree of accuracy), the fluxmeter is a completely relative instrument; measurements are made relative to an arbitrary "zeroed" condition.

This ability to integrate flux over a wide area makes fluxmeters especially useful for characterizing magnets. The major problem encountered when using a gaussmeter for magnet characterization is that the reading obtained is extremely sensitive to the positional relationship between the magnet under test and the measuring probe. While this sensitivity can be reduced by properly fixturing the magnet and probe, it still can be a significant source of error. The other problem with using a gaussmeter for magnet characterization is the issue of what is really being measured. A gaussmeter only measures flux density at a single point in space; it has very little to say about the magnet material characteristics as a whole. By using a fluxmeter, however, it is possible to derive useful information about a material's overall degree of magnetization. Some microprocessor-controlled fluxmeters have options to enter the volume of the magnet under test, so the meter can directly report magnetic properties in a manner independent of magnet size. Because fluxmeter measurements can be performed easily, they are especially useful for performing incoming inspection and other quality control tasks.

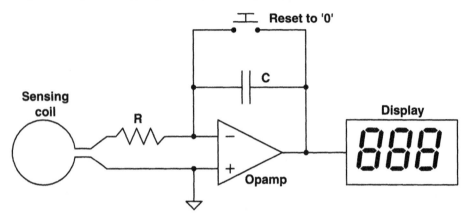

Figure 10-1: Functional block diagram of fluxmeter.

Calibrated Hall-Effect ICs

A calibrated linear Hall-effect IC can be considered a poor man's gaussmeter probe. The principal value of these devices is that, because they can be obtained in the same packages one will ultimately use for sensors in the finished assembly, they can be readily substituted (mechanically) for the ultimately desired IC. This allows one to obtain a measured magnetic response curve from a prototype. Because these devices are much less expensive than gaussmeter probes, they can be viewed as disposable items, and can economically be incorporated into prototype assemblies. Hall-effect IC manufacturers often make these devices available to customers as an aid in developing sensor assemblies.

Polarity probe

It is often useful to be able to distinguish the north pole from the south pole of a magnet. A handheld polarity probe can be used for this purpose. These devices can be bought ready-made, or can be easily built from an Allegro Microsystems UGN3235 dual-output Hall-effect switch and a pair of different colored LEDs, as shown in Figure 10-2.

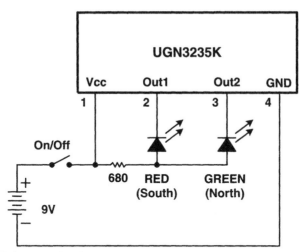

Figure 10-2: Schematic for simple magnetic polarity indicator.

Magnetic View Film

At one point or another in your scientific education you may recall seeing or performing a science experiment in which iron filings are sprinkled on a piece of paper held over a magnet. The filings line up in a way that indicates the magnet's poles and field lines. One drawback of using this method for visualizing magnet pole position is that loose iron filings tend to be messy. The mess-factor, however, has been eliminated by a green plastic film that contains captive ferrous particles. When this film is placed over a magnet, the particles line up accordingly and clearly indicate pole arrangement. This film is especially useful when working with ring magnets, as it allows one to readily see the pole count and position.

Magnetizers and Magnet Conditioners

The ability to magnetize and demagnetize magnets is a useful capability, from both development and production standpoints. The principal tools used to perform these tasks are magnetizers and magnet conditioners.

A simplified schematic of a capacitive-discharge magnetizer is shown in Figure 10-3. This circuit has two operating modes. The first is a "charge" mode in which the capacitor is connected to a charger circuit, which is typically a switch-mode power supply. When the capacitor is charged to a desired voltage level, the switch is then thrown over to "magnetize" mode. The charge stored in the capacitor results in a very high current pulse (often several thousand amperes) that is delivered to the windings in the magnetization fixture. This pulse results in a brief but intense magnetic field that is used to permanently magnetize materials inserted into the fixture. Because the LC circuit formed by the capacitor and magnetizing fixture may oscillate negative, a "flyback" diode may be used in the circuit to protect the capacitor from reverse voltage conditions. Solid-state devices such as SCRs or TRIACs are often used to switch the capacitor to the magnetizing fixture because of their ability to handle very high voltages and currents.

Depending on the value of capacitance, the maximum operating voltage, and the design of the magnetizing fixture, fields up to 50,000–60,000 Oersteds can be developed for a few milliseconds. A field of this magnitude is capable of magnetizing virtually all presently available magnet materials to complete saturation.

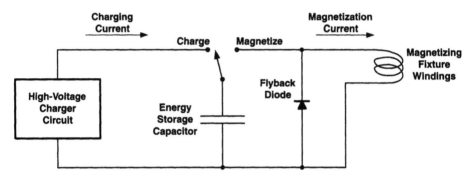

Figure 10-3: Capacitive discharge magnetizer—simplified schematic.

A magnet conditioner is similar to a capacitive discharge magnetizer except that the capacitive discharge is typically allowed to "ring" both positive and negative a number of cycles before it decays to zero. This results in a magnetizing field that alternates polarity, with a reduction in strength on each successive polarity reversal. Applying this kind of damped sinusoidal pulse to a magnet tends to demagnetize it. Partially demagnetizing certain types of magnet materials can be useful because it can reduce their sensitivity to further, unintentional demagnetization as a result of shock and temperature variations.

Because magnetizers and magnet conditioners operate at very high voltage and current levels (typically hundreds of volts and thousands of amperes) , one should not view these kinds of devices as "do-it-yourself" projects. Despite the apparent simplicity of a capacitive-discharge magnetizer, the design of a unit that performs in a safe and reliable manner requires significant amounts of expertise and experience with high-energy electronics and should not be attempted by those without suitable backgrounds. In almost all cases, the best approach to obtaining a magnetizer or magnet conditioning system is to buy one from a reputable manufacturer. When using such a system, follow the manufacturer's instructions for equipment use and maintenance, and exercise due caution when working with these devices.

Helmholtz Coil

It is occasionally useful to be able to generate well-controlled magnetic fields of up to a few hundred gauss for various purposes, principally testing ICs and small sensor assemblies. A small Helmholtz coil can often be used with a power supply for this purpose. An illustration of a Helmholtz coil is shown in Figure 10-4.

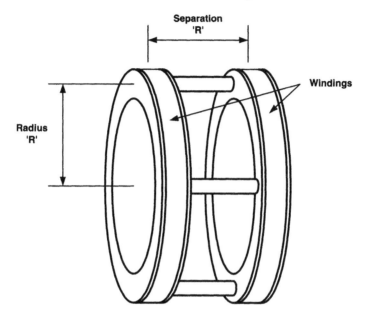

Figure 10-4: Helmholtz coil.

A Helmholtz coil consists of two relatively narrow circular coils that are spaced apart by their radius. The primary feature of a Helmholtz coil is that it produces a highly uniform field over a wide region located between the two windings. The magnetic field developed by an ideal Helmholtz coil, in which the winding cross-section of the two coils is zero, is given by:

$$B = \frac{\mu_0 NI}{r\left(\dfrac{5}{4}\right)^{3/2}}$$

(**Equation 10-1**)

where N is the number of turns on each sub-coil, I is the current through each coil, and r is both the coil radius and separation.

In addition to generating magnetic fields, Helmholtz coils can be used to measure magnetic fields. A Helmholtz coil is often used as the magnetic sensing coil for a fluxmeter.

10.3 Mechanical Tools

One aspect of magnetic sensor development that might come as a surprise to an electronic engineer new to the field is the mechanical-ness of the endeavor. While most EEs are familiar with DVMs, scopes, and other electronic test equipment, tools for mechanical positioning and measurement often cause a bit of culture shock. Nevertheless, when designing sensors that measure mechanical properties, you need some way of generating those mechanical properties. Here are some examples of tools I have found useful.

Optical Bench

An optical bench is a rigid table with threaded holes drilled and tapped into its top surface on a regularly spaced grid. Various fixtures and devices can then be easily and securely screwed down to the table surface. Optical benches can be used when developing sensors to hold sensors and targets and to build up temporary test stands. Small portable optical benches can be obtained from numerous sources. If a high degree of absolute accuracy isn't needed, however, and a good machine shop is available, it is also possible to get one made relatively inexpensively from a piece of steel or aluminum tool plate. An optical bench allows one to rapidly build (and modify) stable test set-ups on which to perform experiments and test prototypes. A small bench, with a good assortment of hex-drive cap screws and small fixtures, can make many measurement tasks much faster and easier to perform, and is much safer and more pleasant to work with than fixtures improvised from c-clamps and double-sided sticky tape!

Linear Positioning Slides

A linear positioning slide consists of a small table running on a linear track containing ball or roller bearings, and provided with some means to accurately set displacement. This device allows you to move the table back and forth by small, precisely controlled amounts. One application of linear slides is for creating precise, repeatable airgaps between sensors and targets. While smaller units may use a micrometer thimble for positioning, larger ones may use a precision leadscrew and ballnut to drive the table slide assembly. Nearly all linear slides have holes provided for mounting to an optical bench or other assembly.

Rotary Table

Like a micrometer slide, a rotary table provides fine control of positioning, but for angular motion. In a typical rotary table, a control knob is used to rotate a worm gear. The worm gear engages the threads of a mating spur gear mounted beneath the table surface. When the worm gear turns, it causes the table to rotate a small amount. By using this arrangement, it is possible to obtain high ratios between rotations of the control knob and rotation of the table, with reduction ratios of 90:1 and 180:1 common. These high ratios make it possible to accurately rotate a target by as little as a few hundredths of a degree. One disadvantage, however, of rotary tables is that they are not usually designed to be rotated quickly and attempting to drive one at high speed can damage the worm gear drive assembly. Figure 10-5 shows an example of a small rotary table.

Figure 10-5: Example of small (4") manual rotary positioning table.

Calipers and Micrometers

A good set of calipers and micrometers are also worth obtaining for one's tool kit. These measuring instruments allow one to easily make dimensional measurements down to 0.001" or better accuracy. Many of these devices come equipped with digital readouts, which make using them extremely simple, compared to the old-style dial and vernier-reading instruments. No toolbox in a sensor development lab should be without a good set of calipers and micrometers.

Machine Tools

Machine tools such as lathes and milling machines can be invaluable when developing sensor products. First, they allow you to rapidly implement small sensor housings and fixtures to try out your ideas. A second application is for modifying housings and fixtures that you buy from an outside source. The ability to accurately cut a slot, drill a hole, or shave a few thousandths off a tight fit can often save you a trip back to your outside machine shop and several days of time. Because most of the machining operations you are likely to want to do to a sensor are relatively small scale, miniature machine tools are often more than adequate for these purposes, and offer the additional benefits of not taking up a lot of space and being relatively inexpensive. Depending on your organization's capital expenditure guidelines, you may even be able to hide many small machine tools and accessories in the budget as generic "tools," thus avoiding the ugly scenario that often occurs when senior management thinks that their engineering staff is setting up their own machine shop.

Another application for machine tools is as temporary sensor test stands. While it is possible to build custom fixturing to mount and spin the target and to hold the sensor at a specified airgap, I have found that miniature machine tools can often be used to perform this function, at least for smaller-diameter targets. Using a tool such as a lathe or milling machine as a test stand can provide several significant features and advantages. The first is that good machine tools have minimal runout and can maintain a tightly controlled sensor-to-target airgap over the course of an entire rotation, assuming of course that the target is concentric. A second advantage provided by some tools is variable speed control. This makes it easy to vary test speed by the turn of a knob. Yet another feature of machine tools is that they normally incorporate precision linear tables for positioning work-pieces and cutting tools, and can accurately position targets and sensor assemblies. Finally, since machine tools are designed to hold and position variously shaped workplaces, the associated clamping and mounting accessories also allow one to set up a given test with a minimal amount of custom fixturing. Figure 10-6 shows an example of a miniature lathe with variable speed control, holding a target and sensor. The cross-slide allows two axes of position control for the sensor.

Figure 10-6: Example of miniature lathe holding target and sensor.

Because machine tools, even the small ones, can develop enough speed and torque to cause serious injury if misused, always follow the manufacturer's instructions and exercise due caution (e.g., wearing safety glasses, keeping fingers away from moving parts, etc.) when operating these devices.

Environmental Chamber

An environmental chamber provides a convenient way to subject sensor assemblies to both hot and cold environments, to see how and if they work at temperature extremes. A good environmental chamber has a programmable controller that allows target temperatures to be set. In many cases the controller also allows a user to enter temperature profiles, which specify a sequence of conditions the oven will automatically generate.

In most environmental chambers, heat is generated by resistive heating elements. Chambers are typically cooled in one of two ways. Larger (and more expensive) environmental chambers have self-contained refrigeration systems to provide cold conditions, while smaller units often rely on external supplies of liquid nitrogen (LN_2) or liquid carbon dioxide (CO_2) for cooling. Both materials have their advantages and disadvantages. LN_2 is more effective at developing low temperatures than CO_2. LN_2's big disadvantage is its limited shelf-life; it is stored cold ($\approx -196°C$) in a Dewar flask and spontaneously boils away over time. On the other hand, liquid CO_2 is stored under pressure at room temperature in a gas cylinder, and can be stored indefinitely, barring valve leakage. The CO_2 in the tank is not actually cold; it cools when it exits the vessel and boils away. When using either a CO_2 or LN_2-charged environmental chamber, it is important to ensure that the area has adequate ventilation to prevent buildup of these gasses to hazardous levels.

If one uses an environmental chamber frequently, it may be worth buying one with an integral refrigeration unit, as this will eliminate the cost, handling, and storage issues associated with either LN_2 or liquid CO_2.

10.4 Magnetic Simulation Software

Although the fundamental equations describing magnetic fields are straightforward to express, they are not as straightforward to solve except in a very few idealized cases. Closed-form analytic solutions may be either difficult or impossible to find except for the most simple magnetic systems. A magnetic system does not need to become very complex before it becomes practically impossible to find closed-form solutions. For this reason a number of approximation techniques have been developed, but they may not yield satisfactory results for complex systems.

Computer simulation offers one solution to the problem of predicting the performance of a magnetic system without having to actually build it. Several computational techniques have been developed for this purpose, with finite-element analysis (FEA) and boundary-element analysis (BEA) the most commonly implemented. In either case, the geometry of the system, the properties of the materials used, and the characteristics of any "sources" (such as electric currents in coils) are known, modern magnetic simulation software can provide a good estimate of how a magnetic system will behave under a variety of conditions.

The first step in building a simulation model is to define the geometry of the system. A short time ago, one was limited to building two-dimensional approximations, and having to make numerous assumptions about how the behavior of the two-dimensional model would extrapolate into three-dimensional reality. Because of advances in both simulation algorithms and computer speed and memory, it is now possible to directly model three-dimensional systems. Figure 10-7 shows an example of a 3-D model built in Ansoft's MAXWELL simulator.

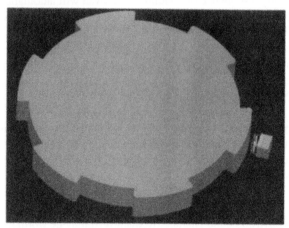

Figure 10-7: 3-D geometric model of geartooth sensor (courtesy of Ansoft, Inc.).

After one defines the model's geometry, the next step is to define the characteristics of the various materials used. Accurate modeling of nonlinear behaviors, such as saturation, is essential to obtaining good results. Improvements in fundamental simulation algorithms have also led to improvements in the area of materials modeling as well.

Once geometry and materials have been defined, the next step is to define sources and boundary conditions. In the case of a static (DC steady-state) magnetic simulation, the two major sources of interest are permanent magnets and electrical currents. In the case of permanent magnets, the magnetization is often defined when one defines an object as being made from a particular material. The direction of magnetization, however, must still be defined. In the case of electrical current, both magnitude and direction also need to be defined. Boundary conditions define how the simulator should behave at the spatial edges of the simulation. Several options are usually available for defining boundary conditions; selecting the proper set is dependent both on the system being modeled and the algorithms used by the simulator.

After geometry, materials, and sources have been defined, the software can begin solving the problem. For the finite element method, the first step is to subdivide the geometry you defined into a collection of smaller subunits, often terahedra or "bricks." This process is commonly called *meshing*. Figure 10-8 shows an example of a mesh. A problem must be meshed for a finite element solution because the method works by defining analytic solutions for the fields within each subunit, and then iteratively solving for a set of solutions that simultaneously satisfy all the relationships among the subunits. Mathematically, this requires solving large sets of simultaneous equations, and is why these types of simulation programs typically require large amounts of memory and a fast processor.

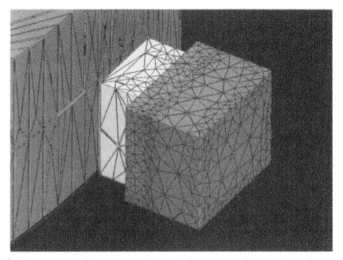

Figure 10-8: Example of finite element mesh (courtesy of Ansoft Inc.).

Finally, after the problem is solved, the results need to be viewed. Presenting the data in the form of raw numbers does not tend to be particularly useful, especially for large problems. For this reason most magnetic simulation packages include some kind of postprocessor, which allows the user to plot data in a number of ways. Some of the more common and useful plots that can be generated by these postprocessors are:

- 2-D plots, magnetic quantity versus position
- Contour maps
- Color-scale maps
- Vector plots

Figure 10-9 shows an example of a vector plot of flux density for the geartooth sensor model.

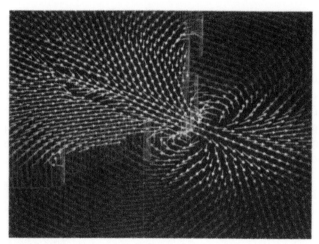

Figure 10-9: Vector plot of magnetic flux density (courtesy of Ansoft Inc.).

Although magnetic simulation tools do not eliminate the need to actually build a sensor and evaluate it, they can vastly speed the process of getting to a design worth prototyping and testing. For the relatively simple magnetic systems used with most Hall-effect sensor assemblies, three-dimensional simulation tools can provide a high degree of accuracy. Because it is much faster to try alternatives out on the computer than in the lab, a larger number of design alternatives can be evaluated, resulting in better product designs. The ability to quickly ask "what if" also can result in much more robust designs, because you can do things such as vary material characteristics and mechanical tolerances to explore design sensitivity issues. The use of magnetic simulation tools should be considered by anyone serious about designing high-quality Hall-effect sensor assemblies.

Appendix A

A Brief Introduction to Magnetics

A.1 Where Magnetic Fields Come From

Magnetic fields result from the motion of electrical charges. The two most common effects that generate magnetic fields are electron spin and moving charges. Some atoms, such as iron, nickel, and cobalt, have imbalances in the total spin of the electrons in their electron shells, and this effect is ultimately responsible for these materials' "magnetic" properties. Moving charges, such as those that form an electrical current, also develop a magnetic field.

The relationships of electric current and magnetic field in empty space under steady-state conditions can be described by:

$$\oint_s \vec{B} \bullet ds = 0 \qquad \textbf{(Equation A.1)}$$

$$\oint_c \vec{B} \bullet dl = \mu_0 I \qquad \textbf{(Equation A.2)}$$

where $\vec{B}$ is magnetic field, I is electrical current density, and μ_0 is the permeability of empty space. Note that $\vec{B}$ is a vector quantity, meaning that it has three separate parts, independently representing field in the x, y, and z directions. Vector notation, although somewhat terse and confusing, is frequently used in electromagnetics because it is vastly less confusing, and more useful, than many alternative representations.

Translated into English, equation A-1 states that the total flux coming through any closed surface, such as a sphere, must equal zero. This means that magnetic poles must come in pairs; for every north pole there must be an south pole of equal magnitude. Another implication of this is that magnetic flux lines always form loops. Equation A-2 states that if you integrate magnetic field along a closed loop (integrating only the field components that are tangent to the loop), the integral will be proportional to the electrical current enclosed by the loop. While the ideas expressed by these equations are simple, using them to predict magnetic fields based on the magnitude and path of an electrical current can be extremely complicated.

Fortunately, it is possible to derive equations that take geometry and current as inputs, and yield magnetic field. One such equation, called the Biot-Savart law, is given by:

$$B = \frac{\mu_0 I}{4\pi} \oint_C \frac{dl \times \vec{R}}{\left|\vec{R}\right|^3}$$ **(Equation A.3)**

The vector $\vec{R}$ defines the distance and direction from the point on the current path one is integrating to the point in space where one wants to know the magnetic field. This geometric relationship is shown in Figure A-1. The three components of the $\vec{R}$ vector are the differences in the x, y, and z coordinates between the two points. The absolute value of $\vec{R}$ is simply the linear distance between the two points. Note that the multiplication is between two vector quantities, and is called the cross-product.

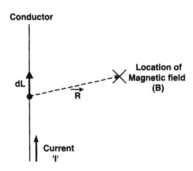

Figure A-1: Geometric interpretation of $\vec{R}$ vector in Biot-Savart integral.

If one has enough skill in evaluating integrals, the Biot-Savart law provides a handy means of predicting magnetic fields from electrical currents. Two of the simpler and more useful results that can be derived from this law are:

The magnetic field around an infinitely long, straight conductor, carrying a current I at a radius of r is given by:

$$B = \frac{\mu_0 I}{2\pi r}$$ **(Equation A.4)**

The magnetic field at the center of a closed circular loop of wire of radius r carrying a current of I is given by:

$$B = \frac{\mu_0 I}{r}$$ **(Equation A.5)**

Illustrations of these two cases are given in Figure A-2.

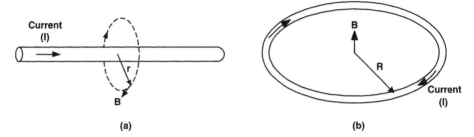

Figure A-2: Magnetic field around wire (a), and inside loop (b).

The above two cases are significant not because you will encounter them in pure form, but because they are useful as quick-and-dirty approximations for many more complex situations. Because the math can get very complicated very fast for problems with even relatively simple geometries, exact analytic solutions for problems in magnetics are both rare and difficult to come by.

A.2 Magnetic Materials

Because all materials contained charged particles, everything exhibits some magnetic properties of one kind or another. For the purposes of this discussion, we will only consider ferromagnetic materials. This group of materials includes iron, nickel, and cobalt, as well as many of their alloys.

Ferromagnetic materials generally are not homogenous down to the atomic level, but consist of clusters of atoms called domains. These domains behave like tiny permanent magnets, each domain having its own north and south pole. In an macroscopically sized, "unmagnetized" piece of a ferromagnetic material, the orientations of all of these domains is random (Figure A-3a), so their individual magnetic fields cancel each other out, and you don't see a net magnetic field. In the case of a permanent magnet, these domains are mostly lined up in the same direction (Figure A-3b), so that the domains' fields reinforce each other.

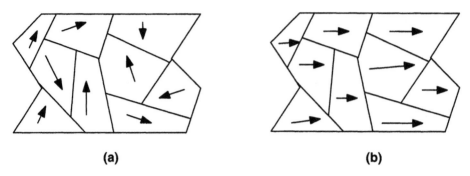

(a)   **(b)**

Figure A-3: Magnetic domains in unmagnetized (a) and magnetized (b) materials.

In a given piece of material, the "easiest" (lowest energy) state for the material to be in is with the domains aligned randomly. To get them to align from a random orientation requires that one apply an external magnetic field. One way of doing this is to wind a coil of wire around the material and apply electrical current to the coil. Because the exact properties of each domain vary somewhat, they don't all flip into alignment at once. As you increase the current in the coil, more and more domains will line up. As more domains line up, you get a stronger magnetic field in the material. At some point, all of the domains are lined up, and applying more current beyond this point doesn't produce a stronger field. At this point the material is said to be saturated. The relationship between magnetizing current and the resulting magnetic field is shown in Figure A-4.
Figure A-4. Relationship between a magnetizing current and resulting field.

Because the amount of magnetizing force you get out of a coil is dependent not only on the current, but also on the number of turns and the geometry, magnetizing force is usually described by a field quantity of its own called magnetic field strength, denoted by the symbol H. H is measured in oersteds. In empty space a magnetizing field of one oersted will result in a magnetic field of one gauss.

In a ferromagnetic material, however, a magnetizing field of one oersted can result in a magnetic field of several thousand gauss. This is because the domains are providing the field, and just have to be coaxed into moving into the desired orientations. This is why you get a much stronger electromagnet by wrapping a coil around a steel nail than you do by just using the coil. The ratio between the change in B versus the change in H ($\delta B/\delta H$) is called relative permeability. Materials such as cold-rolled steels can have relative permeability ranging from 100–10,000. Specialty magnetic alloys such as permalloy or mu-metal can have permeabilities ranging up to 100,000.

Once the domains in a ferromagnetic material have been oriented, in many cases they will want to stay that way when the magnetizing field is removed. To get them to go back to a net unoriented, demagnetized state may require one to reverse the magnetizing field a small amount. If one alternately sweeps a magnetizing field (H) positive and negative, and plots the resulting flux (B) versus magnetizing field (H), one gets a hysteresis diagram like the one shown in Figure A-5.

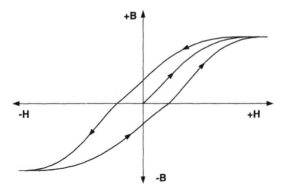

Figure A-5: B-H curve showing magnetic hysteresis.

Some materials, such as soft steels and some ferrites, take very little coaxing to demagnetize. Others, such as samarium-cobalt compounds, take enormous amounts of reverse-magnetizing force to drive them back to a demagnetized state. Materials that strongly resist demagnetization by a reversed magnetizing field are said to have high coercivity. High-coercivity materials are used to make permanent magnets and magnetic recording media. Low-coercivity materials such as steel are used in applications where it is necessary to be able to easily control or vary the amount of flux, such as electromagnets and electrical transformers.

Because one of the more important characteristics of a permanent magnet is how resistant it is to demagnetization, the materials used are often described by only the second quadrant of the hysteresis plot, where magnetizing force (H) is negative and magnetic flux density (B) is positive. This results in the B-H curve commonly used to describe magnet materials, an example of which is shown in Figure A-6.

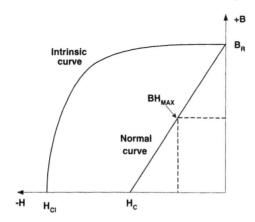

Figure A-6: Second quadrant B-H curves for permanent magnet material.

One may notice that there are two curves in Figure A-6. These are the intrinisic and the normal curves. The relationship between the two curves is given by:

$$B_{normal} = B_{intrinsic} + H$$
(Equation A-6)

The normal curve represents the actual levels of B that appear in the material when subjected to an external magnetizing field of H. The intrinsic curve represents the flux contributed by the magnetic material. There are four points on these curves that are considered important for both characterization of magnetic materials and for magnetic design. These are:

B_r – Remanent induction, which is the flux density (B) present in a closed ring of this material in a saturated state. Measured in gauss.

H_{ci} – Intrinsic coercive force. Used as a measure both of how resistant a material is to demagnetization, and also how much magnetizing force (H) is required to magnetize it to saturation. Measured in oersteds.

H_c – Coercive force. The amount of reverse magnetizing force (H) required to drive the flux density in a closed ring of the saturated material to zero. Also measured in oersteds.

BH_{max} – Maximum energy product. This is the maximum product of B and H along the normal curve. This represents the amount of mechanical work that can be stored as potential energy in the magnet's field. This characterization parameter is therefore very important to people who design electromechanical devices such as motors. The higher a material's BH_{max}, the less of it you need to build a motor of a given power capacity. Measured in mega-gauss-oersteds (MGOe).

If you were to take a gaussmeter and hold it up to the face of a fully "charged" magnet with a B_r specified at 10,000 gauss, you will see considerably less than 10,000 gauss. This is because the B_r figure only appears as actual field when the magnet is in a closed magnetic circuit. In the case of a bar magnet, the empty space the flux must travel through to go from one pole to another effectively "opens" the magnetic circuit. In terms of the B-H curve for that material, this addition of a gap and the resulting reduction in flux (B) corresponds to riding down and left along the normal curve. The point at which the magnetic system rests on the normal curve is called the "operating point." Because the effect of adding a gap to a magnetic system is similar to applying a demagnetizing force (H), the resulting reduction in B is referred to as self-demagnetization. The degree to which a particular magnet experiences this phenomenon is dependent both on the properties of the material of which the magnet is made and also on the magnet's geometry. For a given material, the degree to which self-demagnetization

effects occur is inversely dependent on the magnet's length-to-width ratio (when magnetized along the length). When made of comparable materials, a short, thick magnet
will tend to provide lower flux densities (B) measured at the magnet face than a long
skinny one will.

Calculating the flux density produced at a given point in space by a magnet of a
particular geometry and material can be a nontrivial task, especially for complex geometries. For simple geometries it is possible to obtain closed-form equations that provide
estimates for the cases of the flux density (B) along the axis of both a cylindrical magnet (Equation A-7) and a rectangular magnet (Equation A-8), based on the magnet's
physical dimensions and Br [Dext98a]. Refer to Figure A-7 for an illustration of the
dimensions for these cases.

Cylindrical Magnet:

$$B = \frac{B_r}{2}\left[\frac{d+l}{\sqrt{(d+l)^2+r^2}} - \frac{d}{\sqrt{d^2+r^2}}\right]$$

(Equation A-7)

Rectangular Magnet:

$$B = \frac{B_r}{\pi}\left[\tan^{-1}\left(\frac{d+l}{tw}\sqrt{t^2+w^2+(d+l)^2}\right) - \tan^{-1}\left(\frac{d}{tw}\sqrt{t^2+w^2+d^2}\right)\right]$$

(Equation A-8)

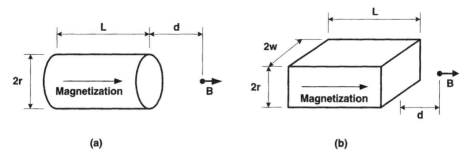

(a)   (b)

Figure A-7: Cylindrical magnet for Equation A-7 and rectangular magnet for Equation A-8.

A.3 Some Permanent Magnet Materials

Several families of materials are commonly used to make permanent magnets. Table A-1 lists the magnetic properties of a few example materials from the more common families.

Table A-1: Common magnet materials and key properties [Dext98b].

Material	BH_{max} (MGOe)	B_r (G)	H_c (Oe)	H_{ci} (Oe)
Ceramic8 (ferrite)	3.5	3850	2950	3050
Alnico5 (cast)	5.5	12800	640	640
Alnico8 (cast)	5.3	8200	1650	1860
SmCo16	16	8300	7500	18000
SmCo28	28	10900	6500	7000
NdFeB31	31	11200	11000	25000
NdFeB44	44	13500	11000	12000

Beyond magnetic properties, magnetic materials also have other characteristics that suit them for particular applications. The following are a few other properties to be considered when deciding to use a particular material.

- Temperature stability
- Service temperature range
- Corrosion resistance
- Hardness and brittleness

Appendix B

Supplier List

This appendix lists various suppliers of sensors, materials, and equipment. This is not an exhaustive listing. Inclusion in this list is neither a recommendation nor an endorsement of the vendor or its products. This information is believed to be accurate at the time this book went to press, but is subject to change without notice.

Hall-Effect Transducers and Sensors:

Allegro Microsystems Inc.
115 Northeast Cutoff, Box 15036
Worcester, MA 01615
Tel: (508) 853-5000
Fax: (508) 853-7861
www.allegromicro.com

Asahi-Kasei Electronics
1-23-7 Nishi-shinjuku,Shinjuku-ku,
Tokyo 160-0023 JAPAN
Tel: +81-3-6911-2800
Fax: +81-3-6911-2815
www.asahi-kasei.co.jp/ake/en

Infineon Technologies Corp.
(U.S. Sales office)
1730 North First St
San Jose, CA 95112
Tel: (408) 501-6000
Fax: (408) 501-2424
www.infineon.com

Lake Shore Cryotronics, Inc.
575 McCorkle Blvd.
Westerville, OH 43082
Tel: (614) 891-2244
Fax: (614) 818-1600
www.lakeshore.com

Melexis Inc.
41 Locke Road
Concord, NH 03301
Tel: (603) 223-2362
Fax: (603) 223-9614
www.melexis.com

Micronas Semiconductors, Inc.
(U.S. sales Office)
8954 Rio San Diego Drive
Suite 106
San Diego, CA 92108
Tel: (619) 683-5500
Fax: (619) 683-3161
www.micronas.com

Optek Technology, Inc.
1645 Wallace Drive
Carrollton, TX 75006
Tel: (972) 323-2200
Fax: (972) 323-2396
www.optekinc.com

Sentron AG
Baarerstrasse 73
CH-6300 Zug
SWITZERLAND
Tel: +41-41-711-21-70
Fax: +41-41-711-21-88
www.sentron.ch

Sypris Test & Measurement, Inc.
6120 Hanging Moss Road
Orlando, FL 32807 USA
Tel: (407) 678-6900
Fax: (407) 677-5765
www.sypris.com

Permanent Magnets:

Adams Magnetic Products Co.
888 Larch Avenue
Elmhurst IL 60126
Tel 800-747-7543
Fax: 800-747-1323
www.adamsmagnetic.com

The Arnold Engineering Co.
300 N. West Street
Marengo, IL 60152
Tel: (800) 545-4578 x427
Fax: (815) 568-2376
www.arnoldmagnetics.com

Dexter Magnetic Materials
1050 Morse Avenue
Elk Grove Village, IL 60007
Tel: (317) 572-1600
Fax: (317) 572-1536
www.dextermag.com

Electrodyne
4188 Taylor Road
Batavia, Ohio 45103
Tel: (513) 732-2822
Fax: (513) 732-6953
www.edyne.com

Kane Magnetics
700 Elk Avenue
Kane, PA 16735
Tel: (814) 837-7000
Fax: (814) 837-9635
www.kanemagnetics.com

Magnequench
9775 Cross Point Boulevard
Suite 100
Indianapolis, IN 46256
Tel: (317) 572-1600
Fax: (317) 572-1536
www.magnequench.com

Toroids and Specialty Magnetic Materials:

Eastern Components, Inc.
11 Union Hill Rd
W. Conshohocken, PA 19428
Tel: (610) 825-8610
Fax: (610) 825-6929
www.eastern-components.com

Fair-Rite Products Corp.
PO Box J, 1 Commercial Row
Wallkill, NY 12589
Tel: (888) 324-7748
Fax (888) 337-7483)
www.fair-rite.com

Magnetics
P.O. Box 11422
Pittsburgh, PA 15238
Tel: (800) 245-3984
Fax: (412) 696-0333
www.mag-inc.com

Magnetic Equipment and Instrumentation:

Lake Shore Cryotronics, Inc.
575 McCorkle Blvd.
Westerville, OH 43082
Tel: (614) 891-2244
Fax: (614) 818-1600
www.lakeshore.com

Magnetic Instrumentation
8431 Castlewood Drive
Indianapolis, IN 46250
Tel: (317) 842-7500
Fax: (317) 849-7600
www.maginst.com

Oersted Technology
24023 NE Shea Lane, Unit #208
Troutdale, OR 97060
Tel: (503) 612-9860
Fax: (503) 692-3518
www.oersted.com

Sypris Test & Measurement, Inc
6120 Hanging Moss Road
Orlando, FL 32807
Tel: (407) 678-6900
Fax: (407) 677-5765
www.sypris.com

Walker LDJ Scientific, Inc.
Rockdale Street
Worcester, MA 01606
Tel: (508) 852-3674
Fax: (508) 856-9931
www.walkerscientific.com

Magnetic Simulation Software:

Ansoft Corporation
225 West Station Square Drive, Suite 200
Pittsburgh, PA 15219
Tel : (412) 261-3200
Fax: (412) 471-9427
www.ansoft.com

Ansys, Inc.
Southpointe
275 Technology Drive
Canonsburg, PA 15317
Tel: (724) 746-3304
Fax: (724) 514-9494
www.ansys.com

Integrated Engineering Software
220-1821 Wellington Street
Winnipeg, Manitoba
CANADA R3H 0G4
Tel : (204) 632-5636
Fax: (204) 633-7780
www.integratedsoft.com

Miscellaneous:

Magne-Rite
E. 17625 Euclid Street
Spokane, WA 99216
Tel: (800) 379-9810
Fax: (509) 922-2586
www.magnerite.com
Film for viewing magnetic fields

Velmex, Inc.
7550 State Route 5 and 20
Bloomfield, NY 14469
Tel: (800) 642-6446
Fax: (585) 657-6153
www.velmex.com
Linear and rotary positioning systems

Sherline Products
3235 Executive Ridge
Vista, CA 92083
Tel: (760) 727-5857
Fax: (760) 727-7857
www.sherline.com
Miniature machine tools and fixturing

Robison Electronics
3480 Sacramento Drive
San Luis Obispo, CA 93401
Tel: (805) 544-8000
Fax: (805) 544-8091
www.robisonelectronics.com
Plastic housings for toroidal current sensors (and other components)

Appendix C

Glossary of Common Terms

Active Area – The surface of a Hall-effect sensor or transducer that is sensitive to magnetic field. For many integrated Hall ICs, the active area is beneath the package surface by a short distance (~0.020" or 0.5 mm), and may only be a few thousandths of an inch square (~0.1 mm × 0.1 mm).

Air Gap – The distance from the face of a magnet to the face of a sensor housing.

Alnico – A permanent magnet material made primarily from aluminum, nickel, and cobalt. Alnico magnets have relatively low energy products, but offer high stability, especially over temperature.

Ampere – MKS unit of electrical current. Equivalent to the passage of $\approx 6 \times 10^{18}$ electrons (or other unit charge carriers) per second.

Ampere-turn – One ampere passing through one turn of wire, typically used to refer to the effective current in a winding in a magnetic circuit such as a toroid.

Ampere-turns/meter – See Amperes/meter.

Amperes/meter (A/m) – MKS unit of magnetic field intensity. If you construct a very long conductive cylinder (ideally infinitely long), and set up a circumferential current flowing around it of one ampere for every meter of cylinder length, you will develop a magnetic field intensity of one ampere/meter inside. Equivalent to about 0.0126 oersteds.

Anisotropic (Oriented) – An anisotropic magnet has a preferred axis of magnetization, and can be magnetized to a much higher degree along this axis than any other. Compare Isotropic.

Auto-Nulling – A group of techniques in which a system's configuration is briefly and periodically changed to measure and eliminate offset errors. After performing an error measurement, the system is then returned to the "operating" configuration and the measured error is subtracted from the signal of interest.

Auto-Zeroing – See Auto-nulling.

Bandwidth – The frequency range over which a device or channel can pass signals. Higher bandwidth is often associated with higher speed.

B-H curve – A curve relating the magnetic flux density (B) in a material to the applied magnetic field intensity (H). Because of magnetization effects, B-H curves often take the form of a hysteresis loop.

B_H – See Hysteresis, Magnetic.

Bipolar – Having two polarities. A signal that can attain either positive or negative values (e.g., ranging from –10 to +10 volts) is said to be bipolar. Contrast Unipolar.

Bipolar Switch – A digital-output Hall-effect sensor whose magnetic switch-points (B_{OP}, B_{RP}) are not necessarily of the same sign. A bipolar switch can therefore be either a normally ON switch, a normally OFF switch, or a latch, depending on manufacturing variations.

Bipolar Transistor – A transistor consisting of a three-layer semiconductor sandwich. Comes in either NPN or PNP versions. Terminals are emitter, base, and collector.

Capacitor – A circuit component consisting of two conductive plates separated by an insulator. Stores energy in the form of an electric field between the conductive plates. Capacitance is measured in farads.

CGS – The metric system of units based on centimeters, grams, and seconds. Gauss and oersted are CGS-derived units. Compare MKS.

Charge Carrier – Something that carries a unit electrical charge. In metals the charge carriers are negatively charged electrons. In semiconductors, positive "holes" can also be charge carriers.

Chopper-Stabilized – See Auto-nulling

Closed-Loop – A description of a system employing negative feedback. In a closed-loop system, the actual output state is monitored and compared to a desired output state. The difference is then fed back to an earlier stage of the system to correct the output towards the desired state. Compare open-loop.

CMOS – Complementary metal-oxide silicon – A technology for building logic circuits that employs both N and P-channel MOS transistors. Allows one to make ICs with lots of transistors that consume small amounts of power.

Coercive Force (Hc) – For a saturated material, the amount of magnetic field intensity (H) required to drive the residual induction (Br) to zero.

Coercivity – A material's degree of resistance to demagnetization. See Coercive Force.

Comparator – An electronic circuit that compares two voltages, and provides either a HIGH output condition if one is greater than the other, and a LOW output condition otherwise.

Concentrator – See Flux Concentrator.

Corner Frequency – The frequency at which the response or gain of a circuit, especially a filter, drops to 0.707 of its nominal value. Also called *–3dB frequency*.

Cunife – A permanent magnet material containing copper, nickel, and iron (Cu-Ni-Fe). One of the few permanent magnet materials that is mechanically ductile.

Curie Temperature (Tc) – The temperature above which a given magnetic material loses its magnetic properties. Not to be confused with the maximum service temperature of a magnet, which is usually considerably lower.

Demagnetization – The process of removing some or all of a magnet's residual induction. Demagnetization can be caused in a number of ways. A magnet conditioner can be used to selectively and deliberately demagnetize a magnet. Heating a magnet beyond its Curie temperature will also cause demagnetization. Finally, the inclusion of an airgap in the path through which a magnet's flux must flow will also demagnetize it (self-demagnetization).

Demagnetization Curve – Quadrant II (upper left quadrant) of a material's B-H curve.

Differential – A difference between two separate measurements. See also Gradient.

Differential Geartooth Sensor – A geartooth sensor that operates based on measuring a gradient in a local magnet field caused by passing gear teeth or other similar ferrous targets.

Digital Hall-Effect Sensor – A Hall-effect sensor that incorporates threshold-sensing circuitry to provide a digital (ON/OFF) output. Common digital Hall-effect sensors include switches, latches, and bipolar switches.

Duty-Cycle – The ratio of ON time to total time. A device operating at a 10% duty cycle is ON 10% of the time, and OFF the remaining 90%. When applied to signals, duty cycle refers to the ratio of time the signal is in a HIGH state compared to total time.

Encoder – A rotation-sensing device using two sensors, providing output signals 90° out of phase. An encoder can be used to determine direction of rotation, as well as speed, and consequently can be used to monitor position.

Ferrite – A family of ceramic compounds that exhibit magnetic properties. Soft ferrites have low coercivity and are used in applications such as current-sensor toroids. Hard ferrites have higher coercivities, making them suitable for use as permanent magnets. Both soft and hard ferrites are mechanically hard and brittle.

Ferrous – Containing iron.

Filter – An electronic circuit designed to pass or attenuate signals of certain specified frequencies. A low-pass filter attenuates signals greater than a particular corner frequency, a high-pass filter attenuates signals below its corner frequency (including DC), while a band-pass filter attenuates signals that fall outside a specified range of frequencies.

Flux – See Magnetic Flux.

Flux Concentrator – A device made from a high-permeability material used to direct magnetic flux in a magnetic circuit.

Flux Density – See Magnetic Flux Density.

Flux Map – A one or two (or maybe even three!) dimensional plot representing magnetic flux density as a function of spatial position. Flux maps of magnetic systems are extremely useful in understanding how to design a compatible sensor.

Fluxmeter – A device that measures changes in magnetic flux, by integrating the voltage induced in a search coil. Useful for characterizing magnets.

Gain – The incremental ratio of an output signal to an input signal. Expressed either in units of output to input (e.g., millivolts/gauss) or as a unitless quantity if the input and output are in the same units (volts/volt). Gain can be used to describe amplifiers, transducers, and many other elements of a system.

Gauss (G) – CGS unit of magnetic flux density. The earth's magnetic field is about 0.5 gauss. 10,000 gauss is equal to one tesla.

Gaussmeter – An instrument for measuring magnetic flux density, typically using a calibrated Hall-effect transducer as the measuring device.

Geartooth Sensor – A sensor used to sense the passage of gear teeth or other similarly shaped features on a ferrous target. A geartooth sensor typically outputs a single pulse as each target feature passes by.

Gradient – A change in a variable vs. a change in position. A magnetic field that varies uniformly by 100 gauss over a distance of 1 cm has a gradient of 100 G/cm.

Head-On – A proximity-sensing mode in which a pole face of a magnet approaches the sensor in a direction normal to the sensor's face.

Helmholtz Coil – An arrangement consisting of two coils of wire of radius R spaced apart coaxially by a distance of R. Helmholtz coils produce a very uniform magnetic field in the region between the coils.

Hysteresis – For an electronic threshold detector, the difference between the input signal levels at which it turns on and at which it turns off.

Hysteresis Loop – The nonoverlapping shape of the B-H curve of a magnetic material. The width of the loop is related to the coercivity of the material, with wider loops indicating higher coercivity.

Hysteresis, Magnetic (B_H) – For a digital Hall-effect sensor, the difference in flux density between the turn-on point (B_{OP}) and the turn-off point (B_{RP}). $B_H = B_{OP} - B_{RP}$.

Impedance – Electrical resistance. Impedance is also used to describe the complex "resistance" of circuits containing inductors and capacitors at non-DC frequencies.

Inductor – A circuit component consisting of one or more windings of wire, sometimes with a ferromagnetic core. Used to store energy in the form of magnetic field. Inductance is measured in henries.

Instrumentation Amplifier (IA, InAmp) – An amplifier having differential inputs and well-controlled gain. Used for precisely amplifying signals from transducers.

Intrinsic Coercive Force (H_{ci}) – A property of a material indicating its resistance to demagnetization. Intrinsic coercive force is the coercive force required to reduce residual induction (B_r) to zero after the material has been magnetized to saturation. On a B-H curve, H_{ci} is the value of H for which B=0 in Quadrant II.

Intrinsic Induction (Bi) – In a magnetic material, the difference between the magnetic induction and a magnetizing force, whether applied externally or developed by the material itself.

Irreversible Losses – Reductions in magnetization in a permanent magnet that can be recovered from by remagnetization. Irreversible losses are often caused by exposure to large magnetic fields, or by heating the magnet.

Isotropic (Unoriented) – An isotropic magnet has no preferred axis of magnetization, and can be magnetized to an equal degree in any direction. Compare Anisotropic.

Keeper – A piece of steel used to form a closed magnetic path around a magnet for shipping and storage. Keepers are used both because they protect a magnet from demagnetization and, in the case of very strong magnets, can make them safer to handle.

Latch – A digital Hall-effect sensor that has B_{OP} and B_{RP} switch points with opposite polarities. Requires a field of one polarity to turn ON, and a field of the opposite polarity to turn OFF.

Leakage Flux – Magnetic flux that escapes from a magnetic circuit.

Linear Hall-Effect Sensor – A Hall-effect sensor that provides an output proportional to sensed magnetic field.

Magnet Conditioner – An instrument for the controlled demagnetization of magnets.

Magnetic Field Intensity (H) – The quantity that defines the ability of an electric current or a magnet to induce a magnetic field (B). Measured in oersteds. In empty space, a magnetic field intensity of 1 oersted will induce a magnetic field (B) of 1 gauss. Also known as Magnetizing Force.

Magnetic Field Strength (H) – See Magnetic Field Intensity

Magnetic Flux (ϕ) – The integral of flux density over area. CGS unit is maxwell, MKS unit is weber. For a uniform normal field over a uniform surface, can be approximated as area times flux density.

Magnetic Flux Density (B) – This is what is commonly called a "magnetic field," and is what you measure when you use a gaussmeter. Analogous to current in an electrical circuit. CGS unit is gauss, MKS unit is tesla.

Magnetic Induction (B) – See Magnetic Flux Density.

Magnetizer – An instrument that can produce intense magnetic fields, used to permanently magnetize or "charge" magnets from their initial unmagnetized state.

Magnetizing Force – See Magnetic Field Intensity.

Maximum Energy Product (BH_{max}) – The point on a material's demagnetization curve where the product of B and H assumes a maximum value. BH_{max} is useful as a measure of how much mechanical potential energy can be "stored" per unit volume of magnet for a given magnet material.

Maximum Service Temperature – The maximum temperature at which a magnet can be exposed without experiencing long-term degradation or structural changes. This temperature is normally lower than, and should not be confused with, the Curie temperature (T_c) for a magnetic material.

Maxwell – CGS unit of magnetic flux. One maxwell is equivalent to a uniform field of 1 gauss normal to an area of 1 cm^2.

MKS – The metric system of units based on meters, kilograms, and seconds. Amperes/meter and tesla are MKS-derived units. Compare CGS.

MOS Transistor – Metal-oxide-semiconductor transistor. A type of transistor commonly used to make digital logic.

Mu-Metal – A high-permeability magnetic alloy consisting primarily of nickel and iron.

Multipole – Having more than one pole. Since all magnets must have at least two poles, when this term is applied to magnets it usually refers to those having more than two poles.

Neodymium-Iron-Boron (NdFeB) – A family of rare-earth compounds used to make very powerful magnets.

Noise – Any signal you aren't interested in. Noise can come from sources internal to a system or from outside sources (interference).

Oersted (Oe) – A CGS unit of magnetic field intensity. One oersted is equivalent to approximately 79.58 ampere/meters.

Offset Error – A type of measurement error in the form of an error signal added or subtracted from the true signal.

Ohm's Law – Voltage V equals resistance R times the current I flowing through it, or $V = IR$. Terms can be rearranged in various ways, solving for current ($I = V/R$) or resistance ($R = V/I$).

Operational Amplifier – An amplifier with very high gain, intended for use in circuits using feedback.

Open-Loop – A description of a system in which the actual condition or state of the output is not monitored and used to adjust or compensate the input to achieve a desired output value. Contrast Closed-Loop.

Operate Point, Magnetic (B_{OP}) – The value of magnetic flux density required to make a digital Hall-effect sensor turn ON.

Operating Point – For a magnetic circuit, the point along the B-H curve that describes the state of the magnetic materials in terms of B and H.

Oriented – See Anisotropic.

Permalloy – A high-permeability ($\mu_r > 10,000$) magnetic alloy made from nickel and iron.

Permeability (μ_r) – Also called *relative* permeability. Permeability expresses the relationship between magnetic flux density (B) and magnetic field intensity (H) in a material. For an "ideal" soft magnetic material, $B = \mu_r H$. Not to be confused with the permeability constant (μ_0), which is a fundamental physical constant.

Power-On Recognition – The ability of a geartooth sensor to detect if a target is present or absent immediately upon being powered-up.

Proximity Sensor – A device that reports the presence or absence of a nearby target.

Quadrature – The condition where two signals are 90° out of phase. Quadrature signals from a pair of sensors are useful for determining rotational direction.

Quiescent Output Voltage (Q_{vo}) –The voltage output by a linear magnetic sensor when no magnetic field is sensed.

Rare-Earth Magnet – A magnet made with a material incorporating a rare-earth element such as neodymium or samarium. "Rare-earth" is a bit of a misnomer, as many rare-earth elements are not all that rare, just difficult to refine into pure form.

Reference Magnet – A magnet that has been calibrated or adjusted so that it produces a known field. Reference magnets are often used as transfer standards for calibration purposes.

Release Point, Magnetic (B_{RP}) – The value of magnetic flux density at which a digital Hall-effect sensor turns OFF.

Remanent Induction (Bd) – The induction (B) that remains in a magnetic material after it has been saturated, and the saturating field removed.

Resistor – A circuit component consisting of two electrodes separated by a poorly conducting material. Resistors are used to limit the amount of current that can flow for a given applied voltage. Resistance is measured in ohms.

Ring Magnet – Magnet formed in the shape of a ring. Often magnetized so that it has a large number of alternating poles along its circumference.

Samarium-Cobalt – A rare-earth compound used to make very strong magnets.

Saturation – For an electronic circuit, saturation is the point at which increasing the input signal does not cause a corresponding increase in output signal. For a magnetic material, saturation is the region on the B-H curve when most of the magnetic domains are aligned, and further increases in magnetic field intensity (H) do not provide corresponding increases in magnetic flux density (B).

Search Coil – A coil of wire used as a transducer element in a fluxmeter. Changes in flux passing through the search coil are transformed into an output voltage.

Self-Demagnetization – The effect that occurs in a permanent magnet when the flux-path is interrupted. Self-demagnetization reduces the flux density in the magnetic circuit.

Semiconductor – A material like silicon or gallium-arsenide that conducts electricity, but not as well as a metal does. Semiconductors are used to make transistors, integrated circuits, and Hall-effect transducers.

Sensitivity – The degree of response of an output parameter to an input parameter. Used to characterize linear Hall sensors (e.g., sensitivity of 2.4 mV/G). See also Gain.

Sensor – A device that measures the state of some environmental parameter. In this book, the term sensor is used to refer to the combination of a transducer with support electronics, as opposed to a simple transducer.

SI – The International System of units. Fundamental units are: meter, kilogram, second, ampere, Kelvin degree (thermodynamic temperature), and candela (luminous intensity).

Slide-By – A proximity-sensing mode in which a pole face of a magnet passes the sensor in a direction parallel to the sensor's face. Compare Head-on.

SOIC – Small Outline IC. A family of surface-mounted IC packages that are very small.

Speed Sensor – A sensor that detects target speed. In the context of Hall-effect sensors, this usually refers to a geartooth sensor or ring-magnet sensor that provides an output pulse for each target feature passing by. Also implies that the edges of the output pulse may or may not accurately track the physical edge of the target; i.e., a particular speed sensor may not be useful as a timing sensor.

Surface-Mount Device (SMD) – An electronic component designed to be soldered to the surface of a printed-circuit board, without requiring holes for leads.

Switch – A digital-output Hall-effect sensor that is normally OFF in the absence of a magnetic field, and only turns on when a minimum flux density is exceeded.

Tempco – See Temperature Coefficient.

Temperature Coefficient – The fraction or percentage by which a property or quantity varies per unit change in temperature. Often expressed in %/° or in ppm/° (parts-per-million/°).

Tesla – An MKS unit of magnetic flux density: 1 tesla = 10,000 gauss.

Thermal Demagnetization – Demagnetization resulting from excessive heating.

Toroid – A donut-shaped flux concentrator often used to make current sensors.

Total Effective Air Gap (TEAG) – The spacing measured from the surface of a magnet to the actual Hall-effect transducer inside any packaging.

Transducer – A device that converts one physical effect into another physical effect. A Hall-effect transducer converts a magnetic field input into a proportional voltage output. A motor is a transducer that converts electricity to mechanical rotation.

TTL – Transistor-Transistor Logic. A popular family of integrated logic circuits, now largely obsolete and superceded by CMOS logic.

Two-Wire Interface – An electrical interface that uses only two wires. Often this means that data is communicated through changes in current, as opposed to changes in voltage.

Unipolar – Having one polarity. A signal that ranges between 0 to +5V is a unipolar signal. Contrast Bipolar.

Unipolar Switch – A switched-output Hall-effect sensor in which both B_{OP} and B_{RP} are of the same polarity. A unipolar switch is ON in the presence of a magnet, and OFF in its absence. Most commercial Hall-effect sensors turn ON when a south pole of a magnet is brought up to their front surface.

Unoriented – See Isotropic.

Vane – A steel flag-shaped target used to interrupt magnetic field between a magnet and a sensor.

Zero-Flux Output Voltage – The voltage output by a linear sensor when no magnetic field is sensed. Also called Quiescent Output Voltage (Q_{vo}).

Appendix D

References and Bibliography

[Avery85] Avery, Grant D., *Ferromagnetic Article Detector with Dual Hall-sensors*, U.S. Patent #4,518,918, May 21, 1985.

[Baltes94] Baltes, H. and Castagnetti, R. "Magnetic Sensors" in S.M. Sze (Ed.*), Semiconductor Sensors* (pp.218-222, 235-237). Wiley & Sons, New York, 1994.

[Bate79] Bate, Robert T. and Erickson, Raymond K. Jr., "Hall-Effect Generator," U.S. Patent # 4,141,026, Feb. 20, 1979.

[Bell] *Hall Generators* (product catalog) F.W. Bell, Orlando, FL.

[Dext98a] *Reference and Design Manual*, pg. 27, Dexter Magnetic Technologies, Hicksville, NY, 1998.

[Dext98b] *Permanent Magnet Catalog*, Dexter Magnetic Technologies, Hicksville, NY, 1998.

[Fowle20] Fowle, Frederick E., *Smithsonian Physical Tables, 7th Edition*, pg. 385, Smithsonian Institution, Washington D.C., 1920.

[Gray84] Gray, Paul R. and Meyer, Robert G., *Analysis and Design of Integrated Circuits, 2nd Edition*, John Wiley and Sons, New York, 1984.

[Kaw90] Kawaji, Hideki and Gilbert, Peter J., *Sensor Having Dual Hall IC, Pole Piece and Magnet*, U.S. Patent # 4,935,698, Jan. 19, 1990.

[Popovic01] Popovic, R.S., Schott, C., Drljaca, P.M., Racz, R., *A New CMOS Hall Angular Position Sensor*, Technisches Messen, tm Vol 6, June 2001, pp.286-291.

[Rams91] Ramsden, Edward A., *Hall Sensor with High-Pass Hall Voltage Filter*, U.S. Patent #4,982,115, Jan. 1, 1991.

[Socl85] Soclof, Sidney, *Analog Integrated Circuits*, p. 485, Prentice-Hall, Englewood Cliffs, NJ, 1985.

[Stauth04] Stauth, J., Dickinson, R., Sauber, J., Engel, R., Pinel, S., *Integrated Current Sensor*, U.S. Patent #6781359, Aug. 24, 2004.

[Sze94] Sze, S.W. (editor), *Semiconductor Sensors*, p.220, John Wiley and Sons, New York, 1994.

[Vig98] Vig, Ravi and Tu, Teri, *Hall-Effect Ferromagnetic Article Proximity Sensor*, U.S. Patent #5,781,005, Jul. 14 1998.

[Vig88] Vig, Ravi, *Two Terminal Multiplexable Sensor*, U.S. Patent # 4,791,311, Dec. 13, 1988.

[Wolfe90] Wolfe, Ronald J. and Hedeen, Larry, *Temperature Stable Proximity Sensor with Sensing of Flux from the Lateral Surface of a Magnet*, U.S. Patent #4,970,463, Nov. 13, 1990.

In addition to the specific references cited above, the following books and publications provide useful material on electronics and magnetics.

Chen, Chih-wen, *Magnetism and Metallurgy of Soft Magnetic Materials*, Dover Publications, New York, 1986.

Cheng, David K., *Fundamentals of Engineering Electromagnetics*, Addison-Wesley Publishing Co., 1993.

Horowitz, Paul and Hill, Winfield, *The Art of Electronics, 2nd Edition*, Cambridge University Press, Cambridge, 1989.

Permanent Magnet Guidelines, Magnetic Materials Producers Association, Chicago, IL, 1998.

Standard Specifications for Permanent Magnet Materials, Magnetic Materials Producers Association, Chicago, IL, 1996.

About the Author

Ed Ramsden, BSEE, has worked with Hall-effect sensors since 1988. His experiences in this area include designing both sensor integrated circuits and assembly-level products, as well as developing novel magnetic processing techniques. He has authored or co-authored more than 30 technical papers and articles and holds six U.S. patents in the areas of electronics and sensor technology. Ed currently resides in Oregon where he designs sensors for the heavy truck industry.

Index

Printed and bound by CPI Group (UK) Ltd, Croydon, CR0 4YY

03/10/2024

01040434-0015